# Pyramidal Neural Networks

Horst Bischof

*Technical University of Vienna*
*Vienna, Austria*

LAWRENCE ERLBAUM ASSOCIATES, PUBLISHERS
1995 Mahwah, New Jersey

Lawrence Erlbaum Associates, Inc., Publishers
10 Industrial Avenue
Mahwah, New Jersey 07430

**Library of Congress Cataloging-in-Publication Data**

Bischof, Horst.
Pyramidal neural networks / Horst Bischof.
p. cm.
Includes bibliographical references and index.
ISBN 0-8058-1913-4 (alk. paper). -- ISBN 0-8058-1914-2 (pbk. : alk. paper)
1. Neural networks (Computer science). I. Title.
QA76.87.B57 1995
006.03'7--dc20 95-23824
CIP

Books published by Lawrence Erlbaum Associates are printed
on acid-free paper, and their bindings are chosen
for strength and durability.

Printed in the United States of America

10 9 8 7 6 5 4 3 2 1

For Marvin

# Contents

# List of Figures

# List of Tables

# Preface

The application of neural networks to image processing and vision tasks is very popular. This book deals also with images and neural networks, however with a different goal. The primary focus is not on the application of neural networks to images but on the transfer of knowledge between image processing/computer vision and neural networks. In order to cope with the complexity inherent in most vision tasks, divide–and–conquer techniques have to be applied that naturally lead to hierarchical structures. It is very important to exploit the capabilities of hierarchies in a systematic way. The approach taken in this book is by considering the similarity between hierarchical neural networks and image pyramids. Throughout this book, all aspects of this similarity are considered, and various results are presented that demonstrate the advantage of this approach. The neural networks resulting from this study have many similarities to image pyramids and are therefore called *pyramidal neural networks*. Each chapter discusses a specific part of pyramidal neural networks and is written as self–contained as possible, so that it can be read independently of the others. At the beginning, each chapter has a short introduction and at the end, a summary highlights the main points. Following aspects of pyramidal neural networks are discussed in detail:

- Chapter 1 provides a motivation for our study. We discuss how hierarchical structures can help to cope with complex problems. Additionally, the goals of the book are outlined.
- After the introductory chapter, the basic concepts of neural networks and image pyramids are introduced in chapter 2. We have also included a review of the major problems that hinder neural networks

to be employed to large scale tasks. In this chapter, a first example demonstrates how to use neural networks in image pyramids.

The following three chapters discuss the three main aspects of pyramidal neural networks in detail, that is, the different kinds of hierarchical structures, the information stored in them, and the types of processing performed on this information. Therefore, the arrangement of the chapters follows naturally.

- In chapter 3, the topology of neural networks and the structure of image pyramids are considered. Rigid regular structures as well as irregular ones are analyzed. We prove that any rigid hierarchical neural network is shift variant. In the case of irregular structures, we develop a new decimation algorithm for irregular pyramids based on Hopfield neural networks. We present an algorithm that uses this new decimation algorithm for image segmentation.
- Chapter 4 discusses the different types of information (numeric and symbolic) to represent in hierarchical structures. We develop a scheme for representing distributed symbolic representations in hierarchies.
- The processing of numeric and symbolic information is analyzed in chapter 5. We develop several novel pyramid algorithms, including sigmoid pyramids, neural curve pyramids, and fuzzy curve pyramids.
- Chapter 6 is devoted to learning in hierarchical structures. Several learning algorithms for grey-level pyramids and their applications are presented. The resulting learning algorithms share close similarity to principal component analysis. The application to feature extraction and image compression is presented.
- Chapter 7 considers learning from another viewpoint. We demonstrate how hierarchies can help in making the learning problem more tractable. The discussion involves decision trees as well as bump-tree structures.
- Finally, chapter 8 summarizes the book and discusses further points of research. In particular, ideas on focusing attention in hierarchical structures are presented.

## Acknowledgments

A large part of this book is my PhD thesis, which was done at the Department for Pattern Recognition and Image Processing (PRIP) at the Technical University Vienna. Many people have contributed to the work presented here. I am indebted to my thesis supervisor Walter Kropatsch who has always shared his ideas with me and has demonstrated to me what it means to work scientifically. His ideas and suggestions have influenced the book to a large extent. Many thanks to Kurt Hornik for a lot of discussions about the subjects in this book. Many thanks also to whole staff at PRIP, who have always supported my work and ideas. I would also like to thank my students who have contributed to the work: Karin Hraby, Stephan Barth, Elmar Thurner, Guido Horracher, Johannes Gartus, Alexander Pucandl, Thomas Nemec, and Thomas Melzer. The work presented on Voronoi Pyramids has been done in collaboration with Etienne Bertin from the University in Grenoble; I hope that we can continue this fruitful collaboration in the future. During writing my thesis, I spent 3 months in Germany at the Institute of Neuroinformatics at Bochum. I thank Professor von Seelen for providing me the possibility to work in his lab, and I thank the staff of the lab for the productive atmosphere.

Special thanks to my family, Sabine and Marvin. They had a hard time during writing the thesis as well as the book, but showed always patience, even when I did not have time for them. I hope Marvin that the time will come when you understand why I did not have time for you. But all "Ballis" that you find in this book are dedicated to you. Finally many thanks to my parents, who have made my studies possible and who always supported my decisions.

# Chapter 1

# Introduction

A large amount of information about the world we live in is supplied by our visual system. Humans perform the task of vision effortlessly, without being aware of this complex process. The goal of computer vision (image understanding) is to build computers that are able to "see," a goal that has not been reached (Kropatsch, 1991).

One major source of inspiration for computer vision is the human visual system. Feldman and Ballard (1982) concluded that *massively parallel models* are the only biological plausible ones, because they satisfy the "100 step rule." That is, because most neurons compute at a maximum rate of 1000 Hz, and because simple perceptual phenomena occur in 100 milliseconds, biologically plausible algorithms can require no more than 100 steps.

Another remarkable fact is the *robustness* of the human visual system. For example, it can operate in a variety of lighting conditions, it can tolerate a lot of noise or damage of components (neurons); it can cope with occlusions; and so on.

Additionally to the above mentioned properties, the human visual system is also quite *adaptive*. That means that the visual system can change its behavior due to different environmental conditions. It is now widely accepted (e.g., Cowan (1987)) that not the whole structure of the brain is genetically determined. After birth, the structure is modified through experience and environmental conditions (e.g., Singer (1986)).

Parallelism, robustness, and adaptivity are also often attributed to artificial neural networks (e.g., Hinton (1989); Pao (1989); Rumelhart & McClelland (1986)), a fact that makes them interesting for computer vision. *Artificial neural networks* (henceforth called *neural networks*), also called *connectionist systems* (Smolensky, 1990) or *parallel distributed processing* (Rumelhart and McClelland, 1986) models, are inspired by the architecture of the biological nervous system. A general principle found in all neural network models is that information is processed in a large number of simple, yet highly interconnected processing elements, called units.

The renewed interest (since the mid–1980s) in neural networks is mainly due to the discovery of new powerful learning algorithms and advances in computer technology (parallel hardware, VLSI technology). In a large number of application examples, the success of neural networks for pattern recognition and computer vision has been demonstrated (e.g., Bischof & Pinz (1990); Bischof, Schneider & Pinz (1992b); Lehky & Sejnowski (1988); Qian & Sejnowski (1988); Roth (1990)).

The use of neural networks in a variety of applications (not only pattern recognition) has been justified by several theorems that prove the strength of neural networks. See Hornik (1991b) for an overview of such results. A particular proof that is often cited (Hornik, 1991a) shows that certain types of networks (three-layer feed forward networks, with bounded and non constant update functions in the hidden units) can approximate any reasonable (Borel) measurable function (and its derivative) to any desired degree of accuracy, if a sufficient number of hidden units is provided.

Another proof by White (1990) shows that these functions are also learnable. The drawback with these results is that they are not constructive, that is, they state that it is possible but do not tell us how. The proofs also do not tell us anything about the efficiency (i.e., space and time requirements) of the resulting neural networks.

Despite these interesting results there are also several deficiencies of current neural network models. One of the main problems is *learning time*

(Hinton, 1989). For networks up to a few thousand units, current models need several hours or even days of learning time. A question closely related to learning is *scaling*, that is, how does the time and/or space requirements change as the problem is scaled up in size. For a popular learning algorithm (back-propagation), it was estimated (Hinton, 1989) that it scales like $\mathcal{O}(N^3)$; $N$ is the number of weights. Another problem closely related to learning is *generalization* i.e., how good is the performance of the neural network on examples on which it has not been trained. There is empirical evidence that neural networks generalize well, however, there have been no formal proofs so far.

The deficiencies of neural networks become especially apparent when a complex problem (e.g., vision) is attacked. In this book, I argue that most of the problems are due to the use of unstructured (e.g., fully connected) networks and that they can be avoided by *modular* and *hierarchically* structured networks. The particular advantage offered by these structures is that complex problems can be decomposed into smaller subproblems that are easier to handle in terms of smaller networks.

Let us put this in a more formal setting:

If a computational process $P$ is defined as the application of some available analysis models $M$ to a set of input data $D$ in order to to solve the problem ($P = (D, M)$), then modularity is a mechanism that allows a computational process $P$ to be divided into a finite sequence of subproblems

$$P_i = \{(D_i, M_i) | D_i \subseteq D, M_i \subseteq M\}, i = 1 \dots n$$

such that the resulting complexity $C = \sum_i (|D_i| \times |M_i|)$ is less than the original complexity ($|D| \times |M|$). $P_i$ might be applied sequentially or parallel, the only constraint is that they must solve the original problem $P$ and that they must operate on the same data $D$ (or a subset of it $D_i$).

Of course, also modularization itself has a cost. Finding an appropriate way to decompose $P$ into $P_i$ may require the application of complex strategies.

The paper by Jacobs, Jordan, & Barto (1990) gives an excellent overview about the advantages of modular structures in terms of:

1. Learning speed
2. Generalization
3. Representation
4. Hardware constraints

The papers (Thurner, 1994; Thurner & Bischof, 1994) provide a systematic overview of different models of modularity in pattern recognition. Hierarchical structures can be obtained from modular ones by recursively applying modularization (i.e., modularizing the modules), which gives a hierarchy of task decompositions.

The advantages of hierarchical structures in the domain of vision have been pointed out by various researchers e.g., (Kropatsch, 1991; Levialdi, 1986; Rosenfeld, 1984; Tsotsos, 1990; Uhr, 1987). Of particular interest in this respect are the papers of Tsotsos (1987; 1990; 1991; 1992), where he applied complexity level analysis to vision, in order to ensure that the basic space and performance constraints of human vision are satisfied. Tsotsos (1990) showed that visual search is an NP-complete problem (requiring an exponential number of steps in the worst case). However, visual search can be reshaped into tractability (polynomial number of steps) through approximation and optimizing the resources devoted to visual processing. He used a "maximum power/minimum cost" principle to rank the many architectures.

The key constraints from this analysis are (see Tsotsos, 1991):

1. parallelism of a sufficiently high degree;
2. hierarchical organization through the abstraction of prototypical visual knowledge in order to cut the search time at least logarithmically;
3. localization of receptive fields, noting that the physical world is spatiotemporally localized and that objects and events, and their physical characteristics, are not arbitrarily spread over time and space;

4. not all visual stimuli require all possible parameter types for interpretation. Separable, logical maps permit selection of individual maps as required;

5. hierarchical abstraction of the input token arrays so as to maintain semantic content yet reduce the number of retinotopic elements;

6. visual parameters at high resolution cannot be directly accessed;they must be obtained by tuning of computing units and through the input abstraction hierarchy;

7. predictions for the overall configurations of the visual system in terms of lower bounds on the size and the number of maps and upper bounds on the required degree of parallelism; and

8. an inhibitory attentional beam.

From this analysis, we conclude that a *hierarchical* structure is *necessary* for a computer vision system in order to deal with the complexity. Indeed hierarchical structures are found in the human visual system (HVS); (see Hubel & Wiesel, 1986; Zeki, 1993). Also in the area of neural networks, hierarchical structures have been used. A typical example is the Neocognitron of Fukushima (1984a; 1984b). A more recent example for the use of hierarchical structures can be found in the work of LeCun (Le Cun, Boser, et.al., 1990; Le Cun, Matan, et.al., 1990a; Le Cun, Matan, et.al., 1990b). A widely used hierarchical structure in the field of image processing are image pyramids (Rosenfeld, 1984). Image pyramids have been studied in great detail in the pattern recognition and computer vision community (e.g., Jolion & Rosenfeld, 1994; Kropatsch 1991; Levialdi, 1986). A lot of algorithms need only $\mathcal{O}(\log N)$ parallel steps ($N$ is the image diameter) on a pyramid, whereas they require $\mathcal{O}(N^2)$ steps without a pyramid.

Figure 1.1 depicts how these three systems are interrelated. The similarity between the Neocognitron and image pyramids has been pointed out by Yamaguchi (Yamaguchi, 1990; Yamaguchi and Kropatsch, 1990). The similarity of image pyramids and the human visual system has been discussed by Rosenfeld (1987), and the similarity between the human visual

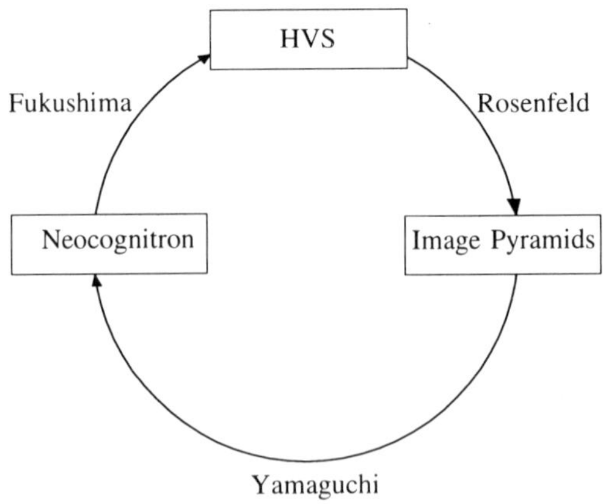


Figure 1.1: Relationship between human visual system, image pyramids, and the Neocognitron

system and the Neocognitron is evident because the Neocognitron is designed to mimic some experimental findings of the visual system (Fukushima, 1984a).

## 1.1 The Goals of This Book

The major goal of this volume is to provide an in depth analysis of hierarchical structured neural networks. In order to achieve this goal, we exploit the similarities and the differences to image pyramids. This comparison can be used to transfer knowledge between these two fields. We show how this knowledge can be used to develop new or enhance existing algorithms for image segmentation, image compression, learning, and attention. The overall goal is the development of a common framework for cellular processing structures.

Working in different areas one has to ask about the advantages gained for each of the areas. Here is a list of advantages offered for neural networks and image pyramids, respectively:

**Image Pyramids:**

Enhancement by neural network functions, especially learning;
finer grained description;
new algorithms for pyramids.

**Neural Networks:**

knowledge about the structure;
formal framework;
decrease of learning time;
better generalization;
improved scaling characteristics.

Another important question we would like to answer is what advantages do neural networks offer compared to conventional schemes like statistics or knowledge–based methods. From a formal point of view, neural networks are equivalent to Turing machines; therefore, they can provide nothing new. However, we argue throughout this book that neural networks provide a different vocabulary for description. In particular, the description offered by neural networks is more "fine grained" compared to other approaches. This offers, on the one hand, the advantage of describing algorithms on a finer level, which gives additional degrees of freedom, therefore making the description more general. On the other hand, it is harder to describe larger structures. In order to do this, we have to define basic building blocks from which we can compose in a hierarchical fashion larger systems. One of the goals of this volume is to offer some of these building blocks for computer vision.

It should be noted that it is not our goal to build biologically plausible models; we are concerned with building computational models. However, we also do not deny some influence (inspiration) from biological nervous systems.

# Chapter 2

# Neural Networks versus Image Pyramids

In this chapter, neural networks (section 2.1) and image pyramids (section 2.2) are introduced in order to define the basic concepts. The goal is to provide a common vocabulary for these two fields. Section 2.3 gives a translation table of the vocabulary used in image pyramids and that used in neural networks. In section 2.4, an example on satellite image (Landsat TM) classification that speeds up considerably by the use of image pyramids is presented. This example serves two purposes:

1. it shows how "conventional" neural algorithms can be applied to image processing and how they compare to classical (statistical) methods; and
2. it demonstrates how much can be gained by the use of image pyramids.

## 2.1 Neural Networks

This section is intended to give a brief introduction and to define the basic vocabulary and notations used throughout this volume. Specific neural network models are discussed in the chapters where they are needed.

It is not the purpose of this section to give an overview about the area of neural networks. Many authors have written excellent books and review articles, therefore we provide only some selected references:

Among the text books are the three Parallel-Distributed-Processing volumes of Rumelhart and McClelland (McClelland & Rumelhart, 1986; McClelland & Rumelhart, 1988; Rumelhart & McClelland 1986) and the book of Pao (1989). A very good introduction is contained in the book of Hertz, Krough, and Palmer (1991). A recent survey about many issues has been done by Haykin (1994). The book by Masters (1993) provides good practical hints, whereas the book of Cichocki and Unbehauen (1993) is written from the view point of electrical engineers.

There are also several collections of articles available. In particular the two books by Anderson (Anderson & Rosenfeld, 1988; Anderson, Pellionisz, & Rosenfeld 1990) include historical articles and give a good overview about the field. Interesting for the large amount of references is also the DARPA neural network study (DARPA (Defense Advanced Research Project Agency), 1988). High quality proceedings papers can be found in the proceedings of the Neural Information Processing Systems (NIPS) Conference (Cowan, Tesauro, & Alspector, 1994; Hanson, Cowan, & Giles, 1993; Lippmann, Moody, & Touretzky, 1991; Moody, Hanson, & Lippmann, 1991; Touretzky, 1989, 1990). Some interesting review articles are Hinton (1989), Carpenter (1989), and Lippmann (1987). The relationships to statistical techniques are pointed out in Ripley (1993).

In the introduction, we argued that neural networks are biologically inspired. What is the difference between brains and computers?[1] Figure 2.1 sketches some of the main differences. In the brain, we have many processors (about $10^{11}$ neurons), executing in parallel a very simple program. In the conventional situation, we have one processor executing a very complicated program sequentially. The connectivity of the brain is far more complex than in computers. Each neuron has $10^3 - 10^4$ connections to other neurons resulting in a total number of $10^{14} - 10^{15}$ connections. Also

[1]When speaking about computers, we mean conventional Von Neumann architectures.

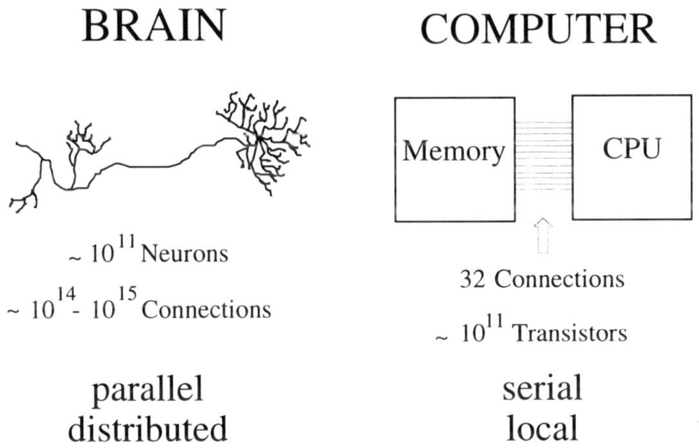


Figure 2.1: Brain versus computer

the kind of information storage is rather different. Brains store information in a highly distributed manner, whereas in computers information is stored locally (i.e., at a fixed memory location). The superiority of the brain in terms of:

- robustness,
- fault tolerance,
- flexibility, and
- speed

is the real motivation for studying neural computation. But it should be noted that we are not doing biological modeling (though such approaches exist; see Kandel & Schwartz, 1985, for a review). We use general organizational principles found in brains to build more powerful computing machines. The connection to neuroscience is not at the level of detailed modeling but at the level of *algorithms and representations* in the sense of Marr (1982). The principles used in neural networks are parallel and distributed processing of information. These features give neural networks a fault–tolerant behavior that is generally referred to as "graceful degradation" (McClelland & Rumelhart, 1986). This means that such systems can operate successfully in noisy environments or when some of the components are damaged (Lippmann, 1987).

Let us now describe what characterizes a neural network in more formal terms. In the sequel we consider:

1. Topology
2. Units
3. Representation
4. Learning

### 2.1.1 Network Topology

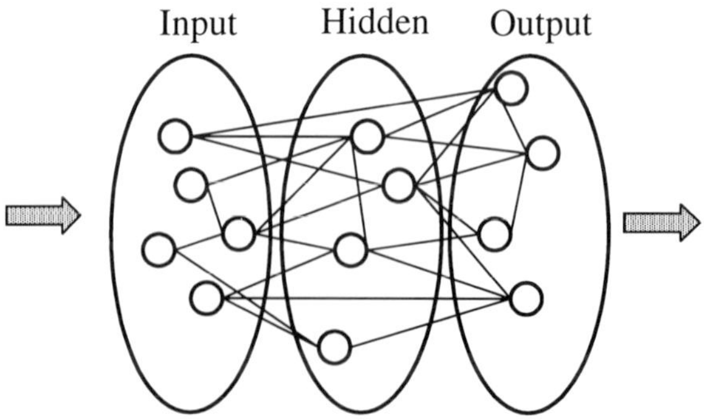


Figure 2.2: Schematic illustration of the network topology

The topology is represented by a directed graph $G = \langle U, E \rangle$ (see Figure 2.2). The nodes $U$ represent the set of *units*, and $E \subseteq U \times U$ is the set of *arcs* in the network. It should be noted that all kinds of directed graphs can be represented by this formalism (except those with multiple arcs between the same units[2]). The set $U$ is partitioned into three sets $I, H, O$ (corresponding to *input, hidden, and output* units) such that $I \cup H \cup O = U, (I \cup O) \cap H = \emptyset$. The predecessors $PR(i)$ of unit $i$ are defined as $PR(i) = \{j | j \in U, \langle j, i \rangle \in E\}$.

Every arc in the network has an associated *weight* $w$, $w : E \mapsto D, D \subseteq \mathbb{R}$. Let the weight from unit $i$ to unit $j$ be $w_{ij} = w(\langle i, j \rangle)$.[3] An arc with an associated weight is called *connection*.

Let us examine some common neural network topologies using our terminology:

- *Two-layer network (perceptron):* The network has only input and output units ($H = \emptyset$). Arcs exist only between input and output units, that is, $E \subseteq (I \times O)$.

- *Multilayer network (multilayer perceptron, feed-forward network):* The hidden units are arranged in layers $H = H_1 \cup \ldots \cup H_n$ and $H_i \cap H_j = \emptyset$ if $i \neq j$. Arcs exist only between adjacent layers (i.e., $E \subseteq (I \times H_1) \cup (H_1 \times H_2) \cup \ldots \cup (H_n \times O)$. If $E = (I \times H_1) \cup (H_1 \times H_2) \cup \ldots \cup (H_n \times O)$, such a network is called fully connected feed–forward network.

- *Fully connected network (Hopfield network):* There exist arcs between all units in the network, that is, $E = U \times U$, and the weights are symmetric, that is, $w_{ij} = w_{ji}$. Sometimes self-connections are not allowed, that is, $E = U \times U - \{\langle i, i \rangle | i \in U\}$.

- *Recurrent network:* The graph $G$ has cycles; this means that there exists at least one path $\langle \langle i_1, i_2 \rangle, \langle i_2, i_3 \rangle, \ldots, \langle i_n, i_1 \rangle \rangle$.

---

[2]If such graphs are required, the definition of $G$ has to be changed to: $G = \langle U, A, f \rangle$ with $U$ the set of units, $E$ the set of arc labels and $f : E \mapsto \langle i, j \rangle, i, j \in U$. Because the basic results are not affected by this extended definition, we choose the simpler notation.

[3]Be careful, some authors use $w_{ji}$ to denote the weight from unit $i$ to unit $j$.

### 2.1.2 Units

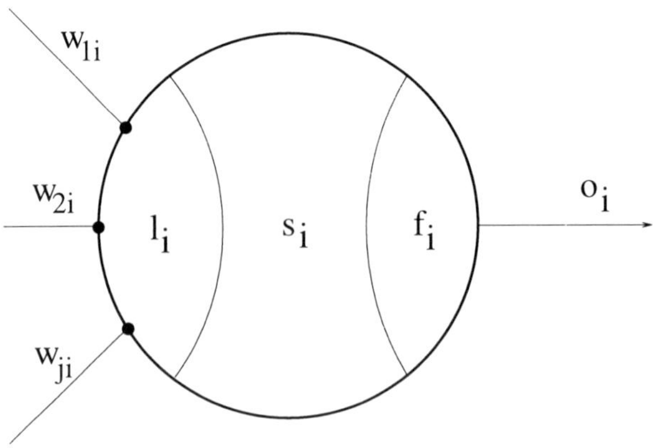


Figure 2.3: Schematic Unit

Processing of information occurs in the units. A schematic diagram of a unit is shown in Figure 2.3. A unit has a *state vector* $s_i \in S \subseteq \mathbb{R}^N$, which describes the internal state of the unit and an *output value* $o_i \in O_i \subseteq \mathbb{R}$, which is sent to other units. The main task of a unit is to compute a new state and a new output value, using the incoming signals (output values of the connected units), the weights of the connections, and the own state vector. Formally this process can be stated as the application of an *update function* $f_i$:

$$f_i : S \times J_i \times W_i \mapsto S \times O_i \tag{2.1}$$

where $w_i \in W_i \subseteq \mathbb{R}^{|PR(i)|}$ is the *weight vector* of unit $i$ (i.e., a vector with all weights of arcs connected to unit $i$) and $j_i \in J_i \subseteq \mathbb{R}^{|PR(i)|}$ is the vector of all incoming signals from other units. Let us call the application of the function $f_i$, *update of unit* $i$.

Typically the update function has the following form:

$$o_i = g(\sum_k w_{ki} j_k + b_i) \tag{2.2}$$

with $g$ either a linear function, a step function, or a sigmoid function (i.e., $g(x) = \frac{1}{1+e^{-x}}$), and $b_i$ is a bias term. Instead of using only linear terms, some models use also high order terms (e.g., terms of the form $w_{ikl}j_kj_l$). These networks are called high order networks (Giles & Maxwell, 1987).

Other forms of update functions are:

$$o_i = g(||w_i - j_i||) \tag{2.3}$$

with $g$ either a linear function in the case of Kohonen feature maps, or radial symmetric functions (e.g., Gaussian) in the case of Radial basis function networks. The other task of a unit is to change the weight vector in order to adapt its behavior. Formally this can be stated in a *learning function* $l_i$:

$$l_i : S \times J_i \times W_i \mapsto W_i \tag{2.4}$$

This is considered in more detail in subsection 2.1.4.

Note that for the update and the learning function, only information locally available at the units is used. The units can therefore operate in parallel and independent of one another.

In summary, a unit $i$ can be represented as a quintuple:

$$i = \langle s_i, o_i, w_i, f_i, l_i \rangle. \tag{2.5}$$

### 2.1.3 Representation

Until now we were only concerned with processing of information inside the network. However, somehow information must enter the network, and the results of the computation must leave the network. How this information is represented at the input units $I$ and at the output unit $O$ is generally referred to as *representation*.

Let the information entering/leaving the network be denoted by $b$ and $z$, respectively. The *basic input* $b \in B$ is a vector with $\alpha$ components. For the sake of simplicity, we assume that the components are from the same set $M$.

$$b \in B = \{M^\alpha | \alpha \in \mathbb{N}\}. \tag{2.6}$$

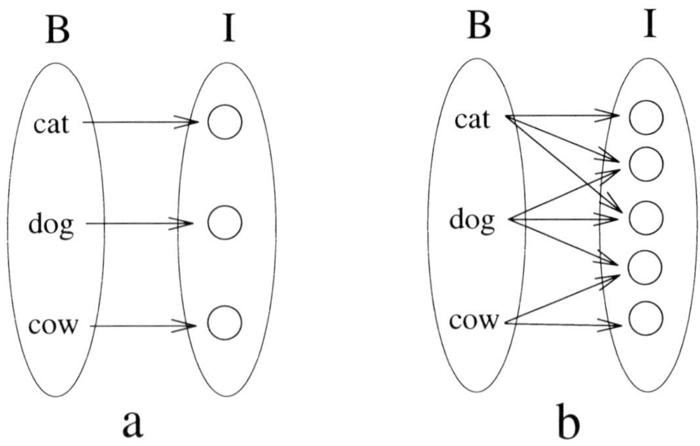


Figure 2.4: Local (a) and distributed (b) representation

The *interpretation vector* $z \in Z = \{N^{\beta} | \beta \in \mathbb{N}\}$ can be defined in a similar manner as the basic input. The *input transformation* $\varphi$ can then be defined as a function from $B$ to the states of the input units.

$$\varphi : B \mapsto S^{|I|} \tag{2.7}$$

where $s \in S^{|I|}$ is the vector of states of all input units. We say $s = \varphi(b)$ is the input representation of $b$. Similarly the *output transformation* $\mu$ can be defined as: $\mu : O_{\mu} \mapsto Z$; where $o_{\mu} \in O_{\mu} \subseteq \mathbb{R}^{|O|}$ is the vector of the output values of all output units. We say that $z = \mu(o_{\mu})$ is represented by $o_{\mu}$.

The neural network literature distinguises two types of representations[4] (Hinton, McClelland, & Rumelhart, 1986; Smolensky, 1990). A representation is called *local* if it dedicates an individual unit to every basic input[5] $b \in B$. When $b$ is to be represented, the unit devoted to $b$ is activated, and all other units have zero activation. A representation is *distributed* if each $b \in B$ is represented by a pattern of activity distributed over many units, and each unit is involved representing many different $b$'s. Figure 2.4 shows a simple example of a local and a distributed representation.

[4]It should be noted that these terms are used differently in the image processing literature.

[5]The same holds also for the interpretation vector $z \in Z$.

Generally speaking a local representation is usually much easier to interpret, but a distributed representation is often more economical in terms of units and more robust against noise (Hinton et al., 1986).

### 2.1.4 Learning

The computations of a neural network depend heavily on the weights $w_{ij}$. Therefore, a mechanism to change (set) the weights is very important; this is generally referred to as *learning*. There are three main categories of learning algorithms:

1. Supervised Learning
2. Reinforcement Learning
3. Unsupervised Learning

A learning algorithm is called *supervised* learning with a teacher) if we provide the network with a "target" output for every input sample of the training set (i.e., the network knows what the correct output value for each output unit should be).

For *reinforcement* learning algorithms (learning with a critic), we provide the network only with a scalar reinforcement signal that is evaluative but not instructive. That is, we do not tell the network what the right output is. In the extreme case, the reinforcement signal is only one bit of information (output is right or wrong).

A learning algorithm is called *unsupervised* (learning without a teacher) when there is no information available to the network about the desired output. In this case, the network must discover itself patterns, features, regularities, correlations, or categories in the input data.

Table 2.1 gives a categorization of many well–known learning algorithms with the corresponding references. Many of the learning algorithms require a special network topology and certain types of units. Most of the

| | |
|---|---|
| supervised | Delta rule Widrow Hoff rule; (Widrow & Stearns, 1985)<br>Perceptron Convergence Procedure (Rosenblatt, 1962)<br>Back-propagation (Le Cun, 1986; Parker, 1985; Rumelhart, Hinton, & Williams, 1986)<br>Recurrent Back-propagation (Almeida, 1988; Pineda, 1987)<br>Artmap (Carpenter, Grossberg & Reynolds, 1991a)<br>LVQ1, LVQ2 (Kohonen, 1989)<br>Boltzmann Machine Learning (Ackley, Hinton, & Sejnowski, 1985) |
| unsupervised | ART1, ART2, ART3, FuzzART (Carpenter & Grossberg, 1987a, 1987b; Carpenter & Grossberg, 1990; Carpenter, Grossberg & Rosen, 1991b)<br>Kohonen feature maps (Kohonen, 1989)<br>Competitive Learning (Rumelhart & Zipser, 1985)<br>Hebb Rule (Hebb, 1949) |
| reinforcement | Associative Reward and Punishing (Barto & Anandan, 1985)<br>Q-Learning (Watkins, 1989)<br>Genetic Algorithms (Goldberg, 1989) |

Table 2.1: Different Learning Algorithms

supervised learning algorithms are based on minimizing a cost function, for example, the summed squared error

$$\mathcal{E} = \sum_t ||y_t - F(x_t)||^2 \tag{2.8}$$

where $y_t$ is the target value for input $x_t$ and $F$ is the function implemented by the network. They differ in how this cost function is minimized. For example, in the case of back-propagation learning (one of the most popular learning algorithms), the cost function is minimized in a gradient descent manner, that is,

$$\Delta w_{ij} = -K \frac{\partial \mathcal{E}}{\partial w_{ij}} \tag{2.9}$$

The major idea of back-propagation is that the partial derivatives can be computed in a recursive way, therefore making the learning algorithm local.

It should be mentioned that learning is not only restricted to changing the weights. Recently some *topology learning* algorithms have been proposed (e.g., Le Cun, Denker, & Solla, 1988; Mozer & Smolensky, 1988). These algorithms change during learning the topology of the network. In Freund, Haberstroh, and Bischof (1992) we proposed a common formalism, based on special types of graph grammars, for weight and topology learning. Also genetic algorithms are used to find "optimal" network topologies (see Albrecht, Reeves, & Steele, 1993 for many articles in this context).

These approaches should be contrasted with the approach in this volume. We try to find a topology by taking into account the problem specification, whereas in topology learning one starts with a "generic" topology that is changed during learning.

#### 2.1.4.1 A Critical View on Neural Learning

This subsection presents several critical ideas and major obstacles of learning algorithms of most of the current neural network approaches. To be concrete we consider computer vision: A vision system can be considered as a different number of maps, for example, visual competences are maps from retinotopic representations or space–time representations to other space–time representations. For any map $m : A \mapsto B$ learning can be considered as a map from $A \times B \mapsto m$. Without loss of generality and to be consistent with the literature, we can call $A$ the stimulus set and $B$ the response set, and we consider the map $m$ as a (visual) behavior. Of course, a behavior may be a composition of maps, but for the sake of simplicity we use behavior and map interchangeably.

Synthesizing maps from examples is an instance of a supervised learning problem. In what follows we always assume supervised learning, though the general arguments apply also in a slightly modified form the other types of learning.

Learning in this respect can be seen from the framework of approximation theory (see Girosi & Poggio, 1990; Poggio & Girosi, 1990). Poggio (Poggio & Girosi, 1990, page 1481) has described this relation as follows: "Approximation theory deals with the problem of approximating or interpolating a continuous, multivariate function $f(X)$ by an approximating function $F(W, X)$ having a fixed number of parameters $W$ belonging to some set $P$ ($X$ and $W$ are real vectors $X = (x_1, x_2, \ldots, x_n)$ and $W = (w_1, w_2, \ldots, w_n)$). For a choice of a specific $F$, the problem is then to find the set of parameters $W$ that provides the best possible approximation of $f$ on the set of "examples". This is the *learning step*.

Seeing learning from the viewpoint of approximation theory provides many fruitful insights (especially on the problems encountered in computer vision), for example, the whole theory of Radial Basis Functions (RBF) relies heavily on this connection. We use this analogy to describe some of the main fundamental problems of learning and of currently used neural network models:

**Local Versus Global Models:** One way to characterize neural networks is by the size of the region in input space that has an influence on a particular unit. If this region is unbounded (i.e., every point in the input space has an influence), we talk of *global models*. A typical global model is a Multilayer Perceptron (MLP). On the other hand, if only a certain (bounded) region of the input space influences a unit, we talk about *local models* (a typical example is a Radial Basis Function network). To illustrate this distinction consider a simple two–class problem in two dimensions as shown in Figure 2.5a. Figure 2.5b shows the decision region of a two layer perceptron that implements this dichotomy, and Figure 2.5c shows the decision regions by a radial basis function networks. Looking at these figures one can see that the representation of the perceptron is more economically. Now consider what happens to the decision regions if we add a new data point as shown in Figure 2.6. In this case the perceptron has to rotate its decision line, which has an effect on the whole input space (i.e., some other points are classified incorrectly), whereas in the case of the radial basis function network this effect is limited to a

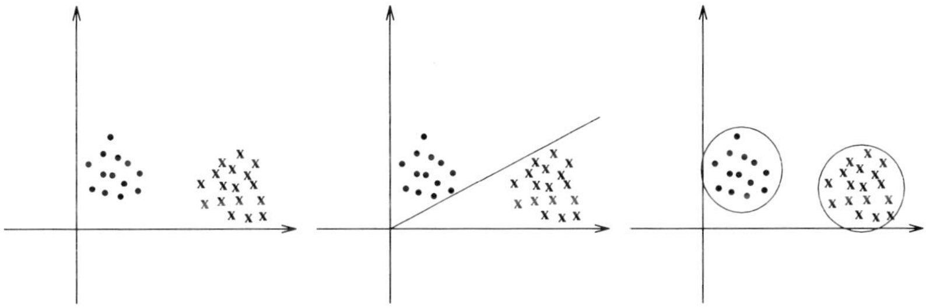

(a) Data points (b) Perceptron decision region (c) RBF decision region

Figure 2.5: A simple example of global versus local models

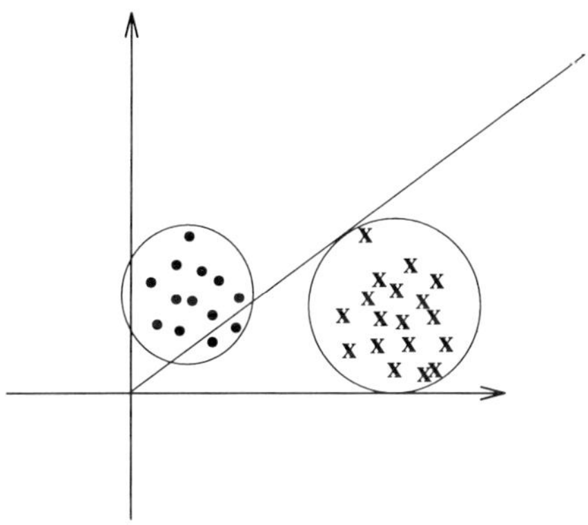

Figure 2.6: Adding one data point and corresponding change in decision regions

much smaller area in input space, which makes the result much more predictable (especially in higher dimensional spaces).

This is also the main reason why radial basis function networks are an order of magnitude faster to train than multilayer perceptrons (Xu, Kryzak, & Yuille, 1994). We will discuss this and other related issues throughout the volume, especially in chapter 7 we make extensive use of these considerations. In various examples and models, we do use multilayer perceptrons, however, one should note that in most cases we use them only in a local region of the input space, that is, global models are restricted in the topology of the network so that they become local ones.

**Curse of Dimensionality:** Many statistical methods as well as neural networks suffer from the so–called *curse of dimensionality*, referring to an exponential increase in resources (e.g., weights) as the dimension of the input space grows. This effect can be best illustrated by a nearest neighbor classifier: Consider a nearest neighbor classifier $C$ operating in a bounded discrete $N$-dimensional space $X$, that is, $C : X \mapsto \Omega = \{\omega_0, \ldots \omega_k\}$, where $\Omega$ is the set of class labels. In this input space, there are $|X|^N$ distinct locations. In order to guarantee good classification performance at least a certain percentage of the locations have to be sampled (i.e., covered by training or reference vectors of the classifier). If we now double the dimension of the input space, that is, $2N$ then we need $\mathcal{O}(|X|^{2N})$ reference vectors, which amounts in an exponential increase in resources of the classifier.

The same effect holds also for radial basis function networks (Haykin, 1994), which have to cover the input space with basis functions. Several techniques have been developed to alleviate this problem (see Poggio & Girosi, 1990; Saha, Christian, Tang, & Wu, 1991). However, the main problem of an exponential increase in resources still remains and seems to be inherent in these local models.

Several authors have claimed that multilayer perceptrons do not suffer from the curse of dimensionality (Barron, 1992). However, one can easily construct problems that need an exponential number of hidden units to achieve a certain classification accuracy. Therefore, this problem seems not to be circumvented by multilayer perceptrons.

The only way to avoid this problem is to operate in a low–dimensional input space. In order to do this, either statistical methods like principal component analysis (Anderson, 1958) have to be employed, or the problem must be decomposed into simpler ones, each of these simple problems requiring only a fraction of available parameters.

**Bias/Variance Dilemma:** A problem closely related to the curse of dimensionality is the *bias/variance dilemma* (Geman, Bienenstock, & Doursat, 1992). This dilemma can be stated in simple terms as follows:

The estimation error (e.g., the error in estimating a probability distribution) can be of two kinds, one related to the structure of the system (i.e., bias) and one related to the variance in estimating the parameters in the system (see Geman, Bienenstock, & Doursat, 1992; Haykin, 1994 for a derivation). A system that can only depart to a certain extent from a given behavior will mainly suffer from an error in bias, whereas a more flexible system will be able to reduce the error in bias, but as a consequence it will have to be punished with a large error due to variance. Only when the size of the training set is allowed to become infinitely large can we hope to eliminate both bias and variance at the same time. Geman claimed that learning of a complex task is essentially impossible without carefully introducing system bias. Needless to say that the introduction of system bias has to be harmless for the problem at hand. Omohundro (1992) claimed following properties to be harmless :

- *Continuity:* The world is to a large extent continuous, and a system should prefer continuous models over discontinuous ones, unless there is contrary data.
- *Sparseness:* The world is sparsely interacting. This says that probable models naturally decompose into components which only directly affect one another in a sparse manner. This results from the fact that physical objects usually only directly affect nearby objects in space and time.
- *Locality:* Models in which the data decomposes into components which are directly affected by only a small number of model components are preferred. This means that only a small number of model parameters affect any individual prediction.

These are some of the most principle obstacles that have to be overcome in order to develop learning systems for large scale tasks. It should be noted that these considerations are also valid for statistical classifiers or symbolic learning techniques. The most promising approach to overcome these principle problems is by inclusion of a priori knowledge (i.e., bias), a strategy that is used in all models introduced in this book.

## 2.2 Image Pyramids

Image pyramids are efficient data and processing structures for digital image analysis (e.g., Jolion & Rosenfeld, 1994; Kropatsch, 1991; Levialdi, 1986; Rosenfeld, 1984; Uhr, 1987). The basic idea is to gradually reduce the spatial resolution of the image and therefore decrease the amount of information. Table 2.2 [6] lists some qualitative consequences of different resolutions.

An image pyramid tries to combine the advantages of high and low resolution. An image pyramid is a collection of images of a single scene at exponentially decreasing resolutions (Tanimoto, 1986). The bottom level of the pyramid is the original image. In the simplest case, each successive level of the pyramid is obtained from the previous level by a filtering operation followed by a sampling operator (Haralick & Shapiro, 1991). More general functions can be used to achieve the desired reduction. We therefore call them *reduction functions*.

| | resolution high | resolution low |
|---|---|---|
| data amount | huge | smaller |
| computing times | long | short |
| details | rich and many | very few |
| overview | bad | good |
| precision | high | low |

Table 2.2: Image Qualities at Different Resolutions

[6] adapted from Kropatsch (1991)

Pyramids have the following supposed advantages (Bister, Cornelis, & Rosenfeld, 1990):

1. reducing the influence of noise by eliminating the importance of details in lower-resolution versions of the image;
2. making the processing independent of the resolution of the regions of interest in the image;
3. converting global features to local ones;
4. reducing computational cost using the divide-and-conquer principle;
5. finding regions of interest for plan-guided analysis at low cost in low resolution images, ignoring irrelevant details.

Image pyramids are closely related to the concept of scale space (Yuille & Poggio, 1986), in which the scale is introduced as an additional continuous dimension (e.g., by smoothing). A pyramid is then a logarithmically sampled version of scale space. Also wavelet transformations (Mallat, 1989) are closely related to image pyramids. For more details on these similarities, see Kropatsch (1991).

There are three important properties that characterize an image pyramid:

1. Structure: for example, neighbors, father–son relations between levels.
2. Contents of a cell: for example, pixel, edge, or more.
3. Processing performed by the cells: for example, filtering.

In the sequel, we discuss these properties with respect to image pyramids. In the subsequent chapters, we analyze these properties in more detail with respect to neural networks.

### 2.2.1 Structure

The structure of a pyramid is determined by the neighbor relations within the levels of the pyramid and by the "father–son" relations between adjacent levels. Figure 2.7a shows a (regular) pyramid. Figure 2.7b shows a particular cell. A cell (if it is not at the base level) has a set of *children (sons)* at the level directly below that provide input to the cell. On the same level, each cell has a set of *neighbor (brothers/sisters)*, and each cell (if it is not the apex of the pyramid) has a set of *parents (fathers)* at the level above.

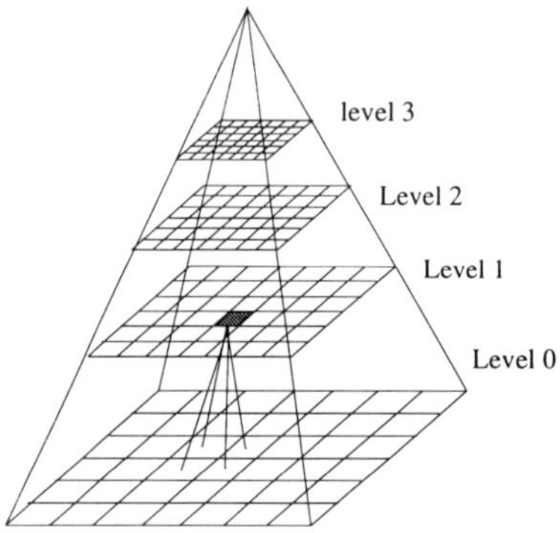


A

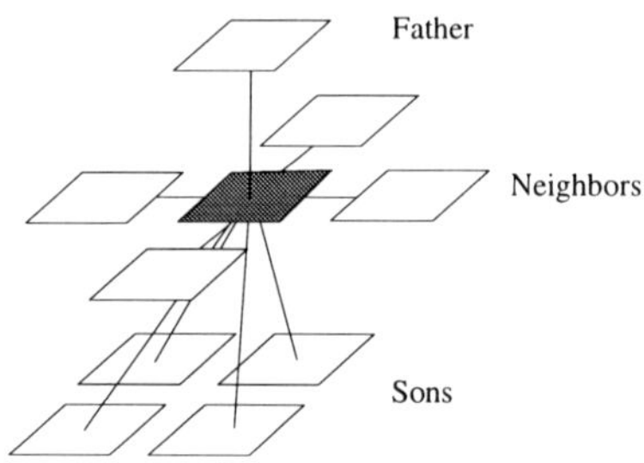


B

Figure 2.7: A pyramid (a) and a particular cell (b)

Therefore, the structure of any pyramid can be described by horizontal and vertical graphs. Each level $i$ of a pyramid can be described by a neighborhood graph $G_i = \langle V_i, A_i \rangle$, where the set of vertices correspond to the pixel of level $i$, and $A_i \subseteq V_i \times V_i$ are the neighborhood relations of the pixels. Two vertices $p, q \in V_i$ are connected in $G_i$ if they are neighbors in the structure.

**Definition 1** *The neighborhood of a vertex $p \in V_i$ is defined by* $\Gamma(p) = \{p\} \cup \{q \in V_i | \langle p, q \rangle \in A_i\}$.

The vertical structure (i.e., the connectivity between the levels) can also be described by a (bipartite) graph: $R_i = \langle (V_i \cup V_{i+1}), L_i \subseteq (V_i \times V_{i+1}) \rangle$. The set of all sons of a cell $q \in V_{i+1}$ is then defined as:

$$SON(q) = \{p | \langle p, q \rangle \in L_i\}. \tag{2.10}$$

In a similar manner, the set of all fathers of a cell $p \in V_i$ is defined as:

$$FATHER(p) = \{q | \langle p, q \rangle \in L_i\}. \tag{2.11}$$

These definitions can be used to define the *receptive field* and the *projective field* of a cell. The receptive field of a cell $p_i \in V_i$ is defined as those cells in the base level ($V_0$) that can influence the cell:

$$\begin{aligned} RF(p_i) &= \{p_0 | p_0 \in V_0, \exists \text{ edges } e_1 = \langle p_0, p_1 \rangle, e_2 = \langle p_1, p_2 \rangle, \ldots, \\ &\quad e_i = \langle p_{i-1}, p_i \rangle, p_1 \in V_1, \ldots, p_i \in V_i\} \end{aligned} \tag{2.12}$$

Similarly the projective field of a cell $p_i \in V_i$ in a given level $V_j, (j > i)$ is the set of cells that are influenced by $p_i$:

$$\begin{aligned} PF(p_i) &= \{p_j | j > i, \exists \text{ edges } e_{i+1} = \langle p_i, p_{i+1} \rangle, e_{i+2} = \langle p_{i+1}, p_{i+2} \rangle, \ldots, \\ &\quad e_j = \langle p_{j-1}, p_j \rangle, p_i \in V_i \ldots, p_j \in V_j\} \end{aligned} \tag{2.13}$$

Any pyramid (regular or irregular) with $n$ levels can be described by $n$ neighborhood graphs and $n-1$ vertical graphs. We distinguish between

- regular structures and
- irregular structures

depending on whether the structural relations are the same for all pyramid cells (except at the boundary) or whether they may vary from cell to cell.

#### 2.2.1.1 Regular Pyramids

Two terms describe the structure of a regular pyramid: the *reduction factor* and the *reduction window*. The reduction factor $r$ determines the rate by which the number of cells decreases from level to level. The reduction window (typically a square $n \times n$) associates to a cell in the higher level a set of cells in the level directly below. The usual notation for describing the structure of regular pyramids is $n \times n/r$. For example, in the classical $2 \times 2/4$ pyramid, a window of $2 \times 2$ forms a new cell of the next lower resolution. Because there is no overlap in this pyramid, the number of cells decreases from level to level by a factor of 4. Figure 2.8 shows a $2 \times 2/4$ pyramid of me.

Figure 2.8: A $2 \times 2/4$ pyramid of me

The quadtree (Samet, 1990) is very similar to the $2 \times 2/4$ pyramid; it differs mainly by the fact that a quadtree has no links to neighbors at the same level. Other pyramid structures with overlapping reduction windows (e.g. $4 \times 4/4, 5 \times 5/4, 2 \times 2/2, 3 \times 3/2$) have also been extensively studied (see Kropatsch (1991) for an overview).

Regular pyramids have one major drawback. Because the structure the same for any image (i.e., it is independent of the contents of the image), these pyramids have problems when the image is shifted, rotated, or differently scaled. Bister et al. (1990) proved that image segmentation algorithms based on regular pyramids are not shift invariant. That is, given an image and a slightly shifted image, the resulting segmentations may differ considerably. In the next chapter, we see that this result can be generalized to a much broader class of problems and structures.

These results provide the necessary motivation for the study of irregular structures that are allowed to adapt to the contents of the image.

#### 2.2.1.2 Irregular Pyramids

Irregular pyramids relax the regularity constraint of regular pyramids. These pyramids operate on a general graph structure instead of the regular neighborhood graph as is the case of regular pyramids.

The motivations for the study of such structures are the following:

1. Shift variance of regular pyramids as proved in Bister et al. (1990), and Bischof and Pinz (1992); see also section 3.1.
2. Perturbations may destroy the regularity of regular pyramids (Meer, Sher, & Rosenfeld, 1988).
3. Biological vision systems are not completely regular (e.g., Kropatsch, 1992).
4. The high cost of large CCD[7] sensors (e.g., $2,000 \times 2,000$ pixels) is mainly due to the fact of their low production yield. The sensors could be produced much cheaper when we allow defective pixels. The resulting sensor geometry would be irregular (see Bederson, 1992; Wallace, Ong, Bederson, & Schwartz, 1993).
5. Generalization to image processing functions to arbitrary pixel geometries (e.g., log-polar image geometries Bederson, 1992; Hartmann, Kräuter, Wiemers, Seidenberg, & Drüe, 1993).

[7] Charge coupled device

When constructing irregular pyramids we would like to preserve as many properties of regular pyramids as possible, that is:

- Only local[8] operations that are independent of every cell (this allows parallelization);
- Bottom up construction of the pyramid, with the number of cells decreasing exponentially (contraction property).

We have to define a procedure that derives $G_{i+1}$ from $G_i$, which is essentially a graph contraction problem. There are two ways to construct irregular pyramids:

1. Parallel graph contraction (Rosenfeld, 1985a)
2. Decimation of the neighborhood graph (Meer, 1989)

Until now algorithms for parallel graph contraction have been devised only for certain kinds of graphs (e.g., trees, Abrahamson, Dadoun, Kirkpatrick, & Przytycka, 1987) therefore we do not consider it further.

Peter Meer (1989) introduced an efficient algorithm for building an irregular pyramid by decimation. Because the pyramid is constructed by a random process, it is called *stochastic pyramid*.

Decimation divides the cells in a pyramid level into two categories: cells that survive form the cells of the next level and cells that do not appear at reduced levels (nonsurvivor). Meer (1989) gave two rules that should be fulfilled by the decimation process. His rules are:

**Definition 2**

1. *Two neighbors at level i cannot survive both;*
2. *a non survivor must be a neighbor of a survivor.*

We call a decimation which satisfies these rules a *valid decimation*. In Meer (1989) it was shown how a decimation can be computed in parallel

[8]Note that this term is used differently in neural networks.

by a stochastic algorithm. It is also worth noting that the rules 1 and 2 are equivalent to saying that the vertices $V_{i+1}$ of the Graph $G_{i+1}$ on level $i+1$ define a maximum independent (vertex) set (MIS) of the graph $G_i = \langle V_i, A_i \rangle$ (Christofides, 1975).

The algorithm for stochastic decimation proceeds in following major steps:

*Algorithm Stochastic Decimation:*

1. Assign uniformly distributed random numbers to the cells $v_i \in V_i$ : $g(v_i) \in (0, M) \subset \mathbb{R}$.
2. Select local maxima as surviving cells: $v_i \in V_{i+1}$ if $g(v_i) > g(v_j)\ \forall v_j \in \Gamma(v_i)$.
3. Fill holes, that is, repeat step 2 as long as there are nonsurviving cells that have no surviving neighbor: a) set $g(v) := 0\ \forall v \in V_i$ : $\Gamma(v) \cap V_{i+1} \neq \emptyset$; b) repeat step 2 if not all $g(v_i) = 0$.
4. Every nonsurviving cell $v_i \in V_i - V_{i+1}$ selects a father $FATHER(v_i) \in V_{i+1} \cap \Gamma(v_i)$. This construction also defines the sons of a surviving node: $SON(v_{i+1}) = \{v_i \in V | FATHER(v_i) = v_{i+1}\}$.
5. Construct the neighborhood graph of the new level: Two surviving vertices $p_{i+1}, q_{i+1} \in V_{i+1}$ become neighbors at level $i+1, p_{i+1} \in \Gamma(q_{i+1})$, if there exists a vertex $p_i \in SON(p_{i+1})$ that has a neighbor in $SON(q_{i+1})$; that is, $\Gamma(p_i) \cap SON(q_{i+1}) \neq \emptyset$.
6. Set $i := i+1$ and repeat steps 1 – 6 with new level until only a single vertex survives, that is, $|V_i| = 1$.

This basic algorithm can be modified in order to take into account the contents of a cell. The concept of the *adaptive pyramid* (Jolion & Montanvert, 1992; Montanvert & Bertolino, 1992) tries first to find significant cells depending on the cell's contents and uses random selection only where the data do not allow a decision. This has been used for image segmentation and connected component analysis in logarithmic time complexity (Montanvert & Bertolino, 1992).

Irregular pyramids offer greater flexibility for the price of less efficient access. In Kropatsch & Montanvert (1992), several properties of irregular pyramids were studied. The main problem of these structures is that the degree of a cell does not remain bounded (for certain configurations, e.g., a tree structure, the degree of a surviving cell increases exponentially). This problem can be circumvented when dealing with planar graphs. It has been proved (Kropatsch & Montanvert, 1992) that the degree of the dual graph does not increase by the decimation process. This property has been used to construct efficient algorithms for the irregular curve pyramid (Kropatsch, Reither, Willersinn, & Wlaschitz, 1993; Kropatsch & Willersinn, 1992a, 1992b, 1992c).

### 2.2.2 Contents

The contents of a pyramid cell is the type of information stored in the cell. Generally speaking, a cell should store information about the region in the image it represents (i.e., its receptive field). In this case, a pyramid cell is considered as an observation window through which a subarea of the base level is observed. Kropatsch (1991) gave two rules that should be satisfied by the representation in a pyramid cell:

1. The region corresponding to that cell in the image space covers the pictorial entity (e.g., primitive image parts, objects, configuration of objects) completely.

2. No smaller cell (cell at lower level) in the pyramid fulfills property (1).

When these rules are satisfied, each pictorial entity is assigned a unique cell in the pyramid in which it should be represented.

In the simplest (and most common) case, a cell stores only one (grey) value. We call such pyramids *grey level* pyramids. In more complicated cases several parameters of general models are stored in a cell (Hartley, 1984). But the basic property that numerical values are stored in a cell remains. Subsequently we call these pyramids *numerical pyramids*.

Besides numerical values it is also possible to store symbolic information in a cell (Kropatsch, 1986). A typical example we consider later in more detail is the $2 \times 2/2$ curve pyramid (Kropatsch, 1986). In this case, we have a finite number of symbols, and a cell stores these symbols or relations among them. We call such a pyramid *symbolic pyramid*.

### 2.2.3 Processing

The main property of processing in a pyramid is that it occurs only locally, that is, every cell computes from the contents of the sons, the brothers, and/or the parents a new value and transmits it to one or more cells of its pyramidal neighborhood. In the bottom–up construction phase, input comes from the sons, but for some algorithms the flow of information is also in the top–down direction (Grosky & Jain, 1986).

The type of operations performed by the cells depends, of course, on the type of the cell's contents. Therefore, we discriminate between:

1. numeric, and
2. symbolic processing

#### 2.2.3.1 Numeric Processing

Three types of filters are commonly used as reduction functions in numeric pyramids.

1. Linear filters
2. Nonlinear filters
3. Morphological filters

**Linear Filters:** One way to linearly filter the image is to convolve it with a spatial filter kernel. Most commonly low pass filters are used. Because a low pass filter suppresses high frequencies in an image, the result can be subsampled with a larger sampling distance without loosing information by sampling. A Gaussian filter kernel has some unique properties (Yuille & Poggio, 1986); it is therefore commonly used in multiresolution representations.

**Nonlinear Filters:** Among the most commonly used nonlinear filters in image pyramids is the *maximum* and *minimum filter* (Blanford & Tanimoto, 1988).

**Morphological Filters:** Haralick, Lin, Lee, and Zhuang (1987) introduced the concept of a morphological pyramid, which is built by morphological operations (Serra, 1982). The input image is first morphologically filtered and then morphologically resampled to the reduced image size.

#### 2.2.3.2 Symbolic Processing

If the contents of a cell consist of symbols or relations among these symbols, we have also to apply symbolic reduction functions. In the general case, a finite state machine (Hopcroft & Ullman, 1979) may be used to perform symbolic reduction. For special types of symbols and/or relations, special reduction functions can be used. For example, Kropatsch (1986) introduced curve relations and a reduction algorithm based on the transitive closure of curve relations. We briefly review the basic steps in building this curve pyramid.

#### 2.2.3.3 $2 \times 2/2$ Curve Pyramid

The basic idea is that linear structures of images are represented by curve relations. A cell of the pyramid is considered as an observation window through which the curve is observed. A single curve intersects this window twice. Only the intersection sides $(N, E, S, W)$ are stored in the cell (i.e., a curve relation). We denote a curve relation by $AB$, where $A, B \in \{N, E, S, W, F\}$ ($F$ is the special end code when the curve ends in a cell).

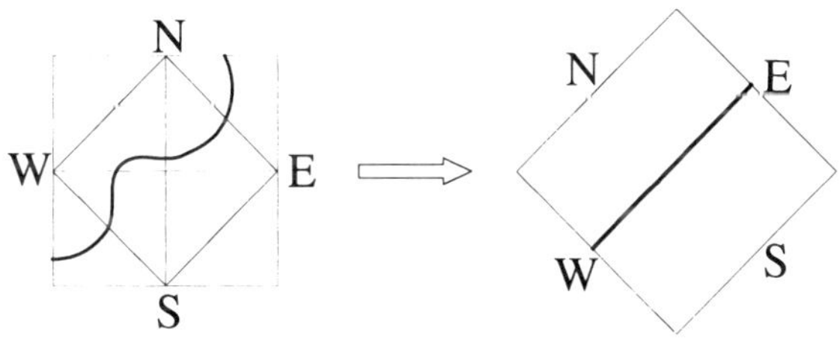


Figure 2.9: A reduction step of the $2 \times 2/2$ Curve Pyramid

The basic routines of building the next level of the pyramid are (Figure 2.9, Figure 2.10):

1. Split – subdivision of the cells contents by introducing a diagonal.
2. Transitive closure – the curve relations of the four son cells are merged by computing the transitive closure of all relations (i.e., $AB, BC \Rightarrow AC$).
3. Merge – the curve relations of the new cell are selected.

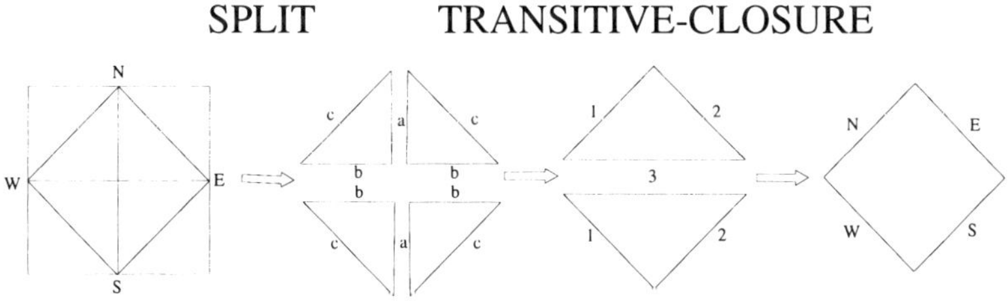


Figure 2.10: Splitting the curve relation by the diagonal

The $2 \times 2/2$ curve pyramid has several interesting properties like the length reduction property, "structural" noise filtering and so forth, which are described in Kropatsch (1986; 1987b).

## 2.3 Translation Table

In the previous sections, we have reviewed the basic concepts of pyramids and neural networks. We have presented some definitions and the vocabulary we use throughout this book. Because the two research areas have introduced different terminologies, Table 2.3 summarizes the equivalent notions of the important concepts in these two fields. This should help one being familiar in one of the two fields to translate his knowledge in the other field. In the sequel, we use the vocabulary for image pyramids when talking about image pyramids and that of neural networks when talking about neural networks. But one should keep in mind that the words can be often used interchangeably.

| Image Pyramids | Neural Networks |
|---|---|
| cell | unit |
| level | layer |
| structure | topology |
| contents of cell | activation of unit |
| bottom-up reduction | activation function |
| parameters of reduction function | weight of arc |

Table 2.3: Translation Table

## 2.4 First Example

The purpose of this example is to show how the state of the art in neural networks can be applied to image processing tasks and how neural networks compare to conventional statistical methods. We show that by the introduction of pyramidal techniques the classification can be considerably sped up. First, we describe the neural network used for classification of Landsat TM images on a pixel by pixel basis (a detailed description can be found in Bischof, Schneider, & Pinz 1992.) Secondly, we show how the networks can be used in a pyramidal framework.

### 2.4.1 Landsat TM Classification

Landsat TM provides high resolution ($30m \times 30m$) satellite images in seven spectral channels. A problem of particular interest is the classification (Kim & Crawford, 1991; Wagner, 1991) of these images (i.e., assigning every pixel a class corresponding to a ground truth value). The data we used for training and testing of the classification accuracy of the neural network were selected from a section ($512 \times 512$ pixels) of a Landsat TM scene of the surroundings of Vienna. In Figure 2.11, channel 4 of this scene is shown.

The aim of the classification was to distinguish between four categories: built-up land, agricultural land, forest, and water. The resulting thematic map is compared with a Gaussian classification (Duda & Hart, 1973). The Gaussian classifier assumes a normal distribution of the data. Preliminary analysis indicated that this requirement was not fulfilled in the case of the four categories. Therefore, these categories were split into 12 sub-categories with spectral data of approximately (not exactly) normal distributions (Wagner, 1991).

Two thematic maps of this scene were prepared by visual classification of the Landsat image, using auxiliary data from maps, aerial photographs, and field work. These two thematic maps showing 4 and 12 land-cover classes were considered to represent the "true" classification of the scene

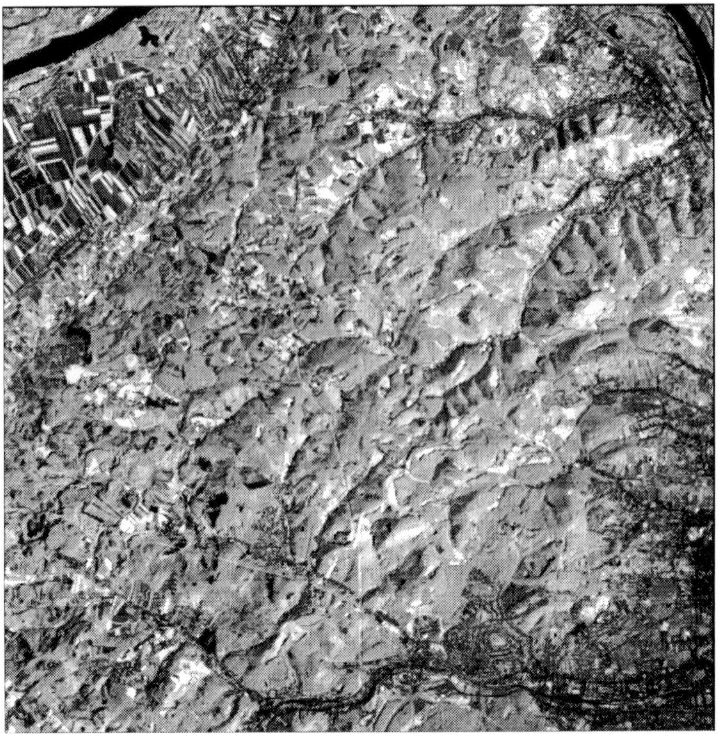

Figure 2.11: Channel 4 of Landsat image

and were used for obtaining both training information and test data for assessment of the accuracy of the automatic classification results (see Figure 2.12). The training set for both the neural network and the Gaussian classifier consisted of 3,000 pixels.

The network used for classification is shown in Figure 2.13. It is a three–layer feed–forward network trained with back-propagation (Rumelhart et al., 1986) and the summed squared error as error measure, which is minimized. The inputs to the network are the seven Landsat TM channels (bands). As input representation one-dimensional coarse coding (Hinton et al., 1986) is used. Coarse coding can be considered as a kind of interpolation. One uses $n$ units to represent the numbers in the interval $[A, B]$. These $n$ units are distributed (usually, but not necessarily) uniformly in this interval. This

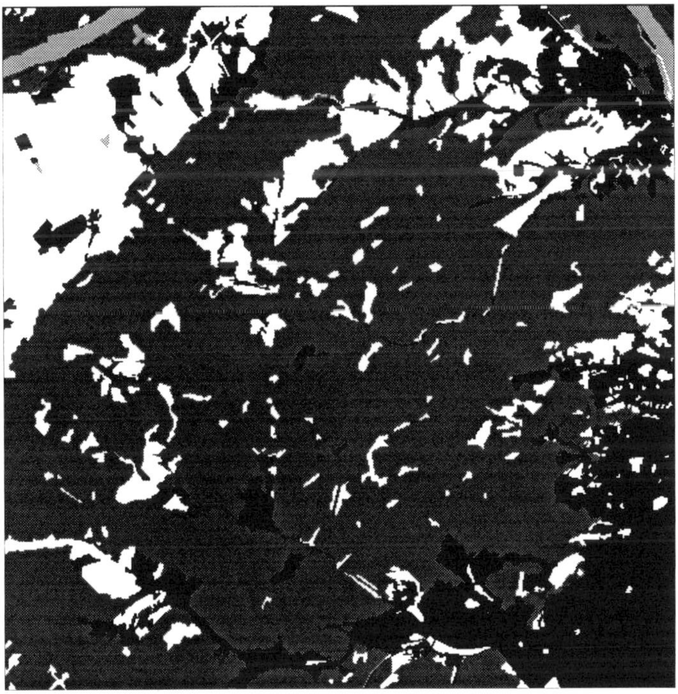

Figure 2.12: Visual classification of Landsat image;black: built-up land; dark gray: forest; light gray: water; white: agricultural area

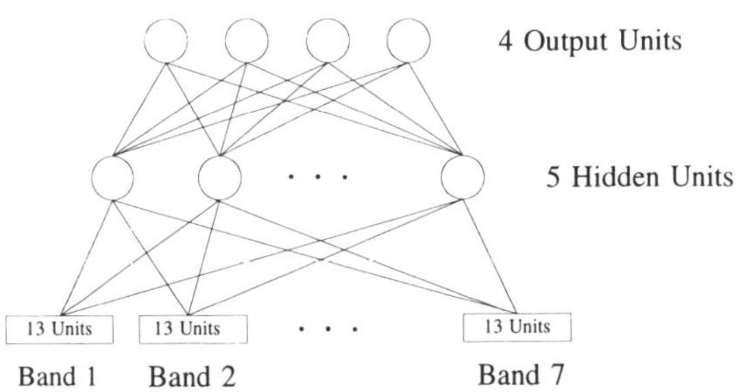


Figure 2.13: Network for classifying Landsat data

means that every unit $i$ is assigned a fixed location $z_i$. When coding a numerical value $m$, the output value of unit $i$ at location $z_i$ is given by a Gaussian response function:

$$o_i(m) = e^{\frac{-(m-z_i)^2}{\sigma^2}} \tag{2.14}$$

where $\sigma$ is a parameter characterizing the width of the curve. We use 13 units for representing one spectral channel resulting in $91 = 7 * 13$ input units. The parameter $\sigma$ is set to 23. Coarse coding is illustrated schematically in Figure 2.14.

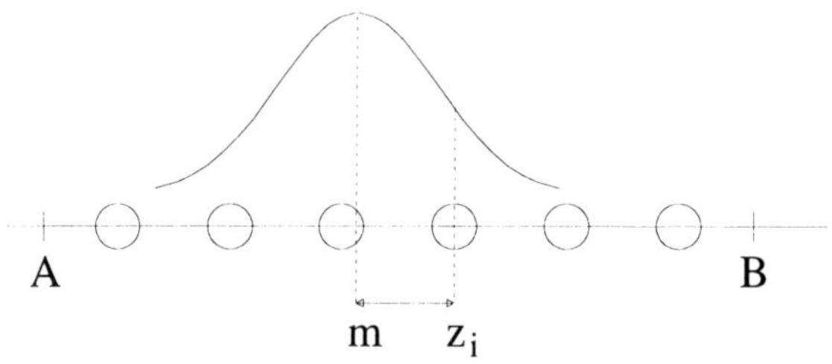


Figure 2.14: Schematic illustration of coarse coding

The output of the classification task is coded locally. Every class is represented by one output unit, which should be fully activated if a pixel of that class is represented at the input, all other units should have zero output value. Therefore, we have four output units (one for each class).

The number of hidden units is a crucial question in neural network design (see Baum & Haussler, 1990; Anshelerich, Amirikian, Lukashin, & Frank-Komenetskii 1989). By experimental studies we found 5 to be a suitable number (Bischof, Schneider, & Pinz, 1992).

The network is trained for 50 epochs (complete presentations of the training set in random order) and achieved 98.1% classification accuracy on the training set. Applying the network to the whole image leads to classification accuracy of 85.9%, which is slightly more than the Gaussian classifier achieved with 84.7% correctly classified pixels (training and classification with 12 classes, merging into 4 classes after classification). Table 2.4 compares the neural network classification with the Gaussian classifier. Figure 2.15 shows the results of the neural network classification.

Neural Network Classification

| | true category | 1 | 2 | 3 | 4 | |
|---|---|---|---|---|---|---|
| 1 | built-up land | **42821** | 980 | 240 | 4898 | 87.5% |
| 2 | forest | 8783 | **145541** | 1091 | 6557 | 89.9% |
| 3 | water | 46 | 29 | **2492** | 37 | 95.7% |
| 4 | agricultural area | 10855 | 3283 | 135 | **34356** | 70.6% |
| | overall accuracy | | | | | **85.9%** |

Maximum Likelihood Classification

| | true category | 1 | 2 | 3 | 4 | |
|---|---|---|---|---|---|---|
| 1 | built-up land | **38286** | 986 | 18 | 9649 | 78.2% |
| 2 | forest | 6824 | **145403** | 181 | 9564 | 89.7% |
| 3 | water | 195 | 5 | **2208** | 196 | 84.7% |
| 4 | agricultural area | 8689 | 3841 | 43 | **36059** | 74.1% |
| | overall accuracy | | | | | **84.7%** |

Table 2.4: Classification Results of Neural Network and Gaussian Classifier

A simple measure of confidence $c$ of the neural network classification can be generated by taking the differences of the output values of the most activated output unit and the second most activated output unit, that is,

$$c = o^{max} - o^{(2)} \tag{2.15}$$

Figure 2.16 shows an image of the confidence measure (white is high confidence, black is low confidence). One can see that the confidence measure

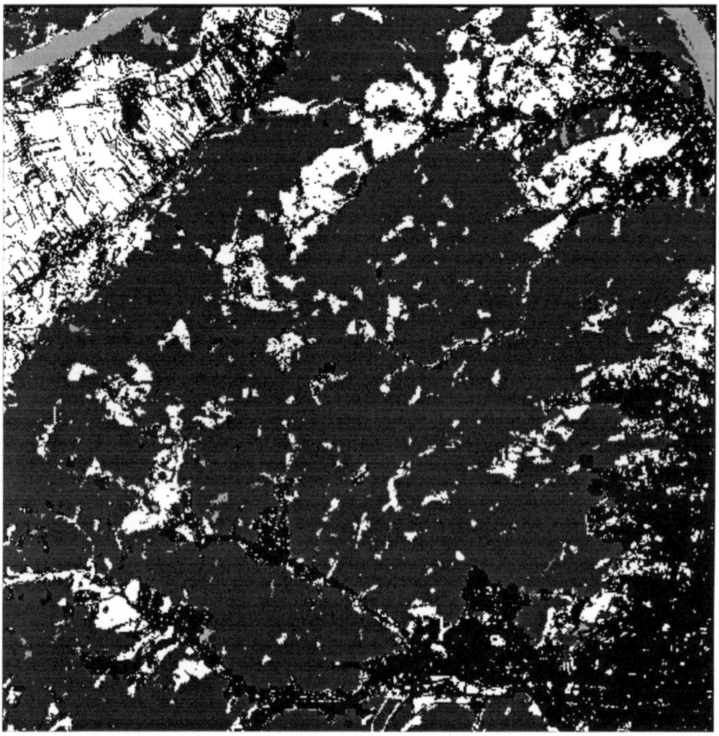

Figure 2.15: Result of neural network classification;black: built-up land; dark gray: forest; light gray: water; white: agricultural area

is low between the borders of adjacent regions. It is very likely that in this case "mixed" pixels (two or more land-cover classes are mapped into one pixel) occur. We use this observation in the next section.

We have extended the basic network architecture to incorporate also textural information in order to decrease the "salt-and-pepper" effect of the classification. In this case, the classification accuracy is raised to 88.1%. We have also experimented with neural networks for postclassificational smoothing, which gave an overall prediction accuracy of 91.0% (for details see Bischof et al. 1992).

In summary, one can say that neural networks give results that are comparable to maximum likelihood classifiers but are not sensitive to the form of

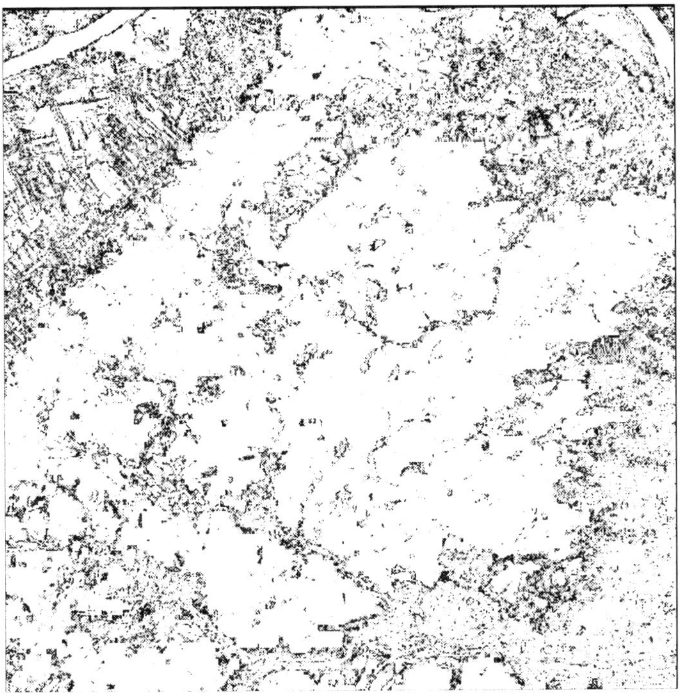

Figure 2.16: Confidence of neural network classification

the underlying probability density function. This gives also the possibility to integrate textural information that improves the classification accuracy of the neural network classifier.

### 2.4.2 Visualization for Network Interpretation

To examine the mode of operation of neural networks, a method for visualizing the weights by so called WV–diagrams (Weight-Visualization diagrams) has been developed (Pinz & Bischof 1990), which has proved extremely useful in the past. The WV–diagrams were modified to the WV–curves, in order to better represent the spectral character of the data. A more general theory of neural network visualization can be found in Bischof and Pinz (1991) and Bischof, Pinz, and Kropatsch (1992a).

In Figure 2.17, the WV–curves for a specific network are shown. Each of the 5 hidden units receives input from 7 spectral channels. Therefore, 7 WV–curves are associated with one hidden unit. A single WV–curve visualizes the weights of the connections from 13 input units; the pixel grey value is coarse coded using 13 units to represent the interval $[0, 255]$. Positive values indicate positive weights from that particular input unit. In the top row of Figure 2.17, the weights from the hidden units to the output units are visualized in a Hinton diagram like manner (Hinton, 1989). There is one square for each output unit, black squares indicating negative weights, the size of the squares corresponding to the magnitude of the weight.

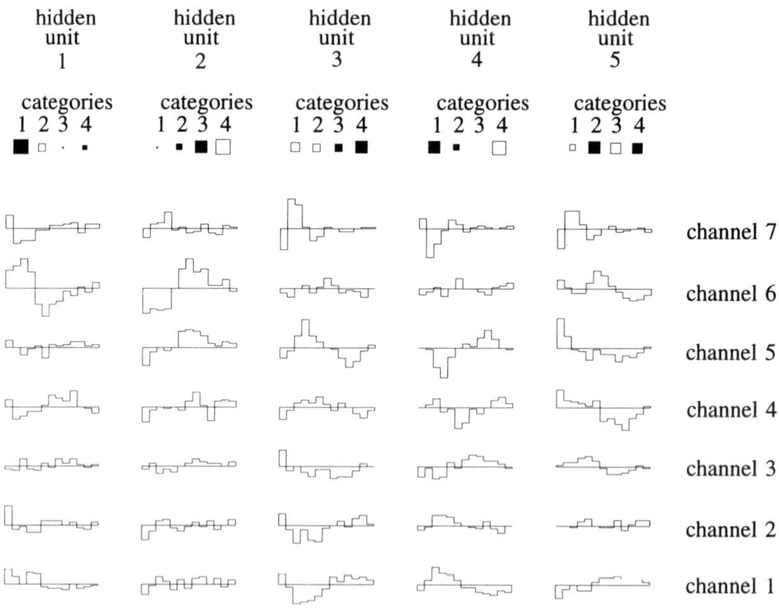


Figure 2.17: WV-curves for a network

Using such diagrams, the network behavior can be interpreted more easily. The internal representation developed by the network and represented by these diagrams can, at least in part, be explained in terms of the reflection curves and temperatures of the various ground cover types. For example, the curve for hidden unit 5 channel 4 (near infrared), which is positively connected to the output units 1 (built-up land) and 3 (water) and negatively connected to the output units 2 (forest) and 4 (agricultural land), indicates a positive contribution for low channel 4 values (no vegetation) and a negative contribution for high channel 4 values (vegetation). For areas without vegetation this unit will therefore favor a decision toward built-up land and/or water, whereas a decision toward forest and/or agricultural land is favored for areas with vegetation. Similarly, the curve for channel 6 (thermal infrared) and hidden unit 2 indicates that low temperatures favor a decision toward category 3 (water), whereas medium temperatures favor a decision towards category 4 (agricultural land).

WV–curves are also useful in feature selection. The weights of the blue channel (no. 1) and the green channel (no. 2) can be seen to be very similar. This means that these two channels are highly correlated so that one of these two channels can be eliminated without degrading the classification accuracy. There are some spectral channels in a few hidden units where the sign of the weights changes for nearby input units (e.g., channel 2 of hidden unit 2). This fact indicates that these channels are hardly used by that hidden unit and may therefore be eliminated.

A similar method for visualization are the so–called OV–curves (Output Visualization curves). In this case, one varies one spectral channel between $[0 \ldots 255]$ and sets all other channels to zero and records the activation of a particular unit. In Figure 2.18 one can see the OV–curves of the output units (four classes) of the network whose weights are shown in Figure 2.17. Using these diagrams an interpretation of the network behavior is even more easier than using WV–curves.

With these and other visualization methods (see Bischof, Pinz, & Kropatsch, 1992), an interpretation of the network behavior is at least partially possible.

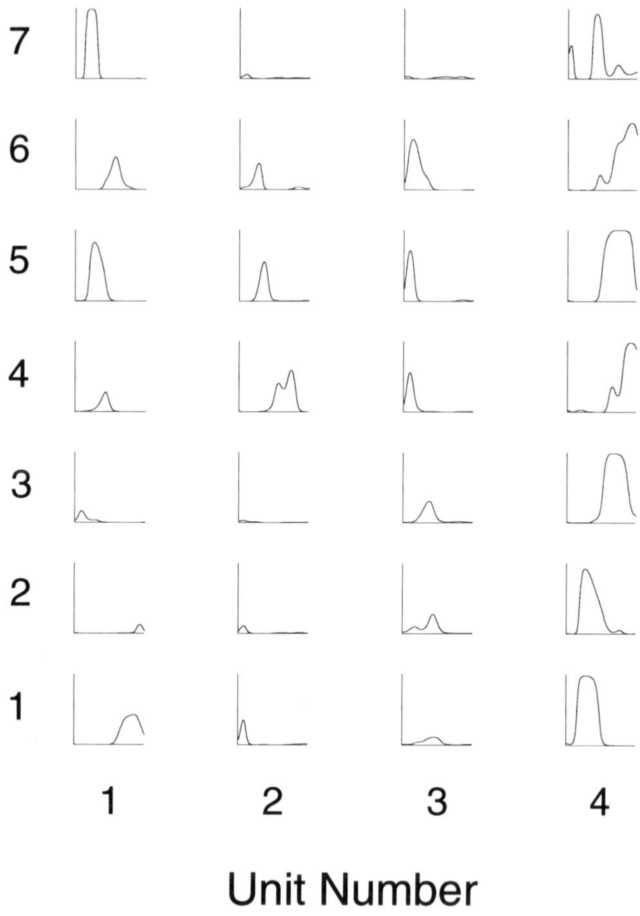


Figure 2.18: OV-curves for the output units of a network

### 2.4.3 Classification in a Pyramid

Looking at the Landsat image (Figure 2.11) and at the classification (Figure 2.15), one can see that it is composed of large homogeneous (i.e., identical land-cover) regions. There is no use in classifying each pixel of such a region individually; it would be sufficient to classify each region only once. In order to do this, we have built a $2 \times 2/4$ pyramid of the Landsat image (actually 7 pyramids for each spectral channel one). The idea is that the network starts classifying the image pixel by pixel at a high level of the pyramid. The classification is accepted if a confidence measure is high. In this case, all pixels of the receptive field on the base level are assigned to this class. Otherwise if the confidence of the classification is low, a network on the next lower level of the pyramid is called. The network classifies the four son pixels. If the confidence of the classification of one of these pixels is low too, the process continues on the next lower level, and so on until the base level is reached.

This process will yield good results if the input image has large homogeneous regions. In this case, networks at lower levels have to be invoked only at the borders of such regions. The critical feature in this process is the confidence measure of the classification. An ideal confidence measure should be high if the pixel is from a homogeneous region, and should be low if several classes have contributed to this pixel value (i.e., mixed pixel). These conditions are equivalent to estimating mixture distributions (Kim & Crawford, 1991). We have studied three kinds of confidence measures:

**Algorithm 1:** We have seen in the previous section that the confidence measure $c$ is sensitive to mixed pixels. Therefore, we can use this measure also in the pyramid. However, a problem might occur if only a few pixels from other classes are present in a large homogeneous region. These pixels get lost by the averaging process and can therefore not be recovered. This can also be interpreted as a certain kind of generalization, which is a common practice in thematic mapping (e.g., usually a classification is smoothed by some majority based filter). Another more severe problem arises when the averaging of two or more classes gives a pixel value that is similar to another class. In this case, large errors occur.

**Algorithm 2:** To deal with this problem we investigated a more complex confidence measure, which makes use of the fact that we are in a multiresolution hierarchy. The process is illustrated in Figure 2.19 for one pixel at level $i$ of the pyramid. Assume that this pixel is classified as $\omega_p$ with confidence $c_p$ (the same confidence measure as previous). The four son pixels on level $i-1$ are classified as $\omega_{p1} \ldots \omega_{p4}$ with the corresponding confidences $c_{p1} \ldots c_{p4}$. A reasonable choice for accepting the classification of level $i$ is when $\omega_{p1} \ldots \omega_{p4}$ are the same as $w_p$ and the confidence of the classification of the son pixels is high. In all other cases, a network at level $i - 2$ must be invoked for some pixels of level $i - 1$.

**Algorithm 3:** The basic idea of this algorithm is that we predict from the low resolution the high resolution information. This is similar to the Laplace pyramid (Burt & Adelson, 1983). We first construct a pyramid, then we classify the highest level. Then we expand this level ($i$) and compare it with level $i - 1$. Whenever there is a large error (greater than a threshold $\theta$), we classify those pixels with the high resolution information. Otherwise we take the classification from the lower resolution. The threshold $\theta$ can be used to control the amount of calculations. For example, if we want only 50% of the pixels to be classified, we can choose at each level a $\theta$ such that this requirement is fulfilled.

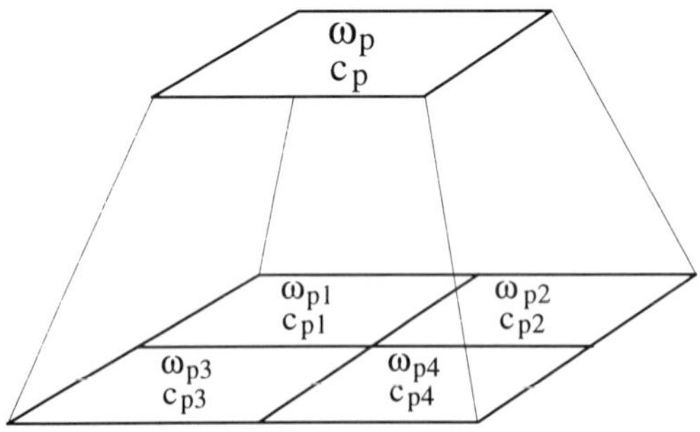


Figure 2.19: Illustration of the complex confidence measure

#### 2.4.3.1 Results

We have tested all three algorithms on the Landsat TM scene and compared it to the visual classification. For all algorithms, we have used the same network that was also used to classify the original image; the process was started for all algorithms at level 4 ($32 \times 32$ pixels) of the pyramid. We tested Algorithms 1 and 2 on a $2 \times 2/4$ pyramid and Algorithm 3 with a $3 \times 3/4$ Laplace pyramid. The threshold of Algorithm 1 was .94. For calculating the error of Algorithm 3, we have used the maximum error of all seven spectral channels. The threshold $\theta$ was set to 15 (i.e., a pixel was reclassified if one of the Laplace images of the spectral channels had a greater absolute value than 15). The results of these experiments are shown in Table 2.5 and in Figure 2.20.

One can see that all three methods provide similar overall classification accuracies (note the raise for forest areas for Algorithms 1 and 2; forest is the largest homogeneous region in the image). Although the bottom level classification (Figure 2.15) has a salt-and-pepper appearance, the results for Algorithms 1 and 2 are much smoother. Using Algorithm 1, small details get lost (as expected). These details are to some degree preserved by Algorithms 2 and 3. From a computational point of view, the original bottom level classification required 262,144 invocations of the neural network, whereas Algorithm 1 required only 42,876 invocations. This is approximately 16% of the computational time of the bottom level classification. Algorithm 2 was more expensive (as expected) and required 70% of the computation of the bottom level classification. Algorithm 3 provides a 50% reduction of computational cost. The costs of building the pyramids can be neglected compared to the classification (approximately 2% of the bottom level classification).

This example demonstrates that a pyramid can considerably improve the performance of neural networks and decrease the computational cost. We have not exploited this model in full detail (e.g., overlapping pyramids, training of the network in the pyramid, etc.). The purpose was to show on a real world example how pyramid techniques can improve neural networks, thus providing a motivation for studying the relationships between these

| | original network | algorithm 1 | algorithm 2 | algorithm 3 |
|---|---|---|---|---|
| built-up land | 87.5% | 84.4% | 90.1% | 88.4% |
| forest | 89.9% | 94.9% | 92.8% | 89.7% |
| water | 95.7% | 89.1% | 93.5% | 95.3% |
| agricultural area | 70.0% | 58.1% | 65.2% | 70.1% |
| overall accuracy | 85.9% | 86.1% | 87.2% | 85.9% |
| time | 100% | 16% | 70% | 50.5% |

Table 2.5: Classification Results of Neural Networks in Image Pyramids

methods in depth, which is done in the next chapters.

## 2.5 Summary

In this section, we have introduced neural networks and image pyramids. We have presented the basic definitions and notations. For neural networks, we have presented several major problems which have to be overcome in order to apply these models to large–scale real world tasks. From this analysis, we have derived with a translation table of terms used in neural networks and image pyramids. We have also seen that there is much similarity between these two fields that has not been fully exploited until now.

At the end of this chapter, we have presented a first example that shows how neural networks may profit from image pyramids. We have presented a neural network that classifies Landsat TM data and compared it to Gaussian maximum likelihood classifications. We have shown how this network can be used in image pyramids, which drastically reduces the processing time required for classification of a whole image.

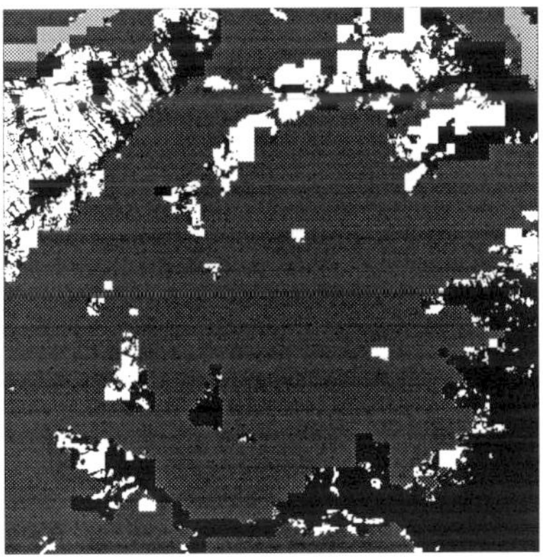

(a) Algorithm 1

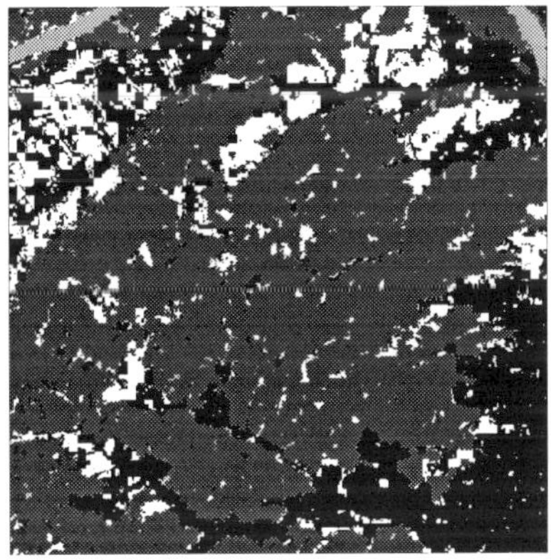

(b) Algorithm 2

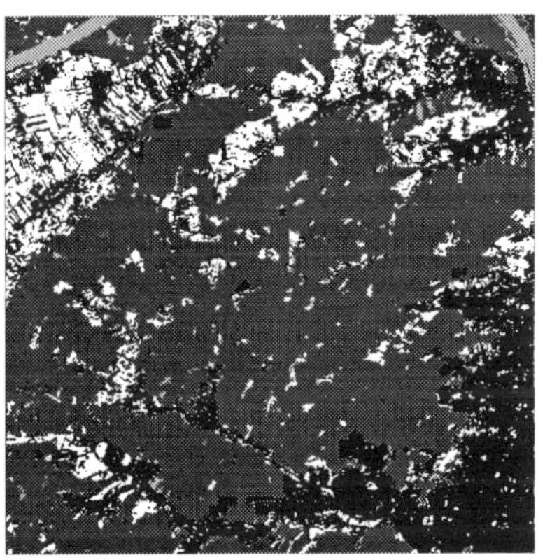

(c) Algorithm 3

Figure 2.20: Results of the three pyramidal classification algorithms;black: built-up land; dark gray: forest; light gray: water; white: agricultural area

# Chapter 3

# Structural Considerations

In section 2.1.1 and 2.2.1, we have seen that both the topology of neural networks and the structure of pyramids can be described by graphs. In the case of neural networks, we have made no restrictions on the graph structure, whereas in the case of pyramids, we have $n-1$ vertical graphs $R_i = \langle V_i \cup V_{i+1}, L_i \subseteq (V_i \times V_{i+1})\rangle$ and $n$ neighborhood graphs $G_i = \langle V_i, A_i\rangle$ ($n$ is the number of levels). These $2n-1$ graphs can, however, be combined in a single graph:

$$G = \langle V = \bigcup_{i=0}^{n-1} V_i, A = A_{n-1} \cup (\bigcup_{i=0}^{n-2} (L_i \cup A_i))\rangle \tag{3.1}$$

It is clear that we can build a neural network topology that is equivalent to the graph G. The converse is true only for certain types of neural networks. For example, Yamaguchi and Kropatsch (1990) showed that the Neocognitron can be simulated on a pyramid structure. Therefore, we conclude: *We can build for any pyramid structure an equivalent neural network topology*. From now on we call neural network topologies that are equivalent to some pyramidal structure *hierarchical neural networks*. We can use this result to transfer some knowledge between these two fields. We would like to demonstrate this by two examples: The first concerns rigid (regular) pyramid structures, where we show that shift variance results hold also for hierarchical neural networks. The second example demonstrates that we can use Hopfield neural networks to construct irregular pyramids.

## 3.1 Shift Variance of Hierarchies

A vision system should be capable of recognizing objects irrespective of their position in the image. Stated more formally, the system should be shift invariant. Sometimes we also would like to have scale and rotation invariance. The local effect of a small scale change or a small rotation can be compared to a local shift. Therefore, if a system is not locally shift invariant, it will also be not locally invariant to scaling and rotation (Bister et al., 1990). For this reason, we concentrate on the shift variance.

There is strong evidence that rigid hierarchical structures are not shift invariant. Two recent results indicate this fact:

1. Bister et al. (1990) showed that pyramid segmentation algorithms based on regular and consequently rigid structures are not shift invariant. That is, the segmentations of the image might differ considerably when the image is shifted by a small amount. The basis of their proof is that due to the sub-sampling in pyramids, a one pixel shift in the image cannot be represented in the pyramid. The conclusion from this observation is that the same image content is represented differently, which results in different segmentations.

2. Barnard and Casasent (1990) showed that the unsupervised version of the Neocognitron which is also a hierarchical structure (Fukushima 1984a, 1984b) is not intrinsically shift invariant. This is an especially interesting result because the Neocognitron has been designed to be shift invariant. The basis of the proof is that if the parameters of the Neocognitron are set such that the output of a particular layer is shift invariant, then the discrimination ability is lost (i.e., two different patterns can no longer be discriminated). On the other hand, if the discrimination ability is preserved, a pattern and a shifted pattern will be classified differently.

These results hold only for certain kinds of structures and processing, subsequently, we show this for a more general class of structures and types of processing. But let us first examine different methods (and their shortcomings) that have been proposed (also for neural networks) to deal with shift variance:

1. Invariant feature spaces like Fourier-transform, Mellin-transform, log-polar transform (Grace & Spann, 1991; Seibert & Waxman, 1989), and many others. But so far invariant feature spaces that can tolerate arbitrary variations have not been found (Hiari & Tsukui, 1990). These transforms are also sensitive against noise.

2. The structure of the network can be designed to be invariant under certain transformations(e.g., Neocognitron [Fukushima, Miyake, & Ito, 1983], high order networks [Giles & Maxwell, 1987]), but this approach suffers mainly from the complexity of the resulting network or its rigid behavior.

3. Training an arbitrary network with a lot of different examples of a pattern under the transformation (Rumelhart & McClelland, 1986). This approach has many shortcomings. First the networks must be huge to learn many patterns. Secondly, if a new object is encountered, the network has to be trained anew with this pattern under all different transformations.

4. Changing the topology of the network for every new input pattern. This approach is used in irregular pyramids (Meer, 1989), which change their connectivity according to the input data. That is, the structure of the pyramid is allowed to change to account for transformational effects.

### 3.1.1 Variance Proof

The hierarchical structure we consider is shown in Figure 3.1. We have an input image at the bottom level (level 0). From this image several local features (e.g., edges, color, texture, etc.) are computed and represented in a retinotopic fashion (level 1). These features are then combined by a hierarchical structure into more global ones (level 2 to level $n$), thereby reducing

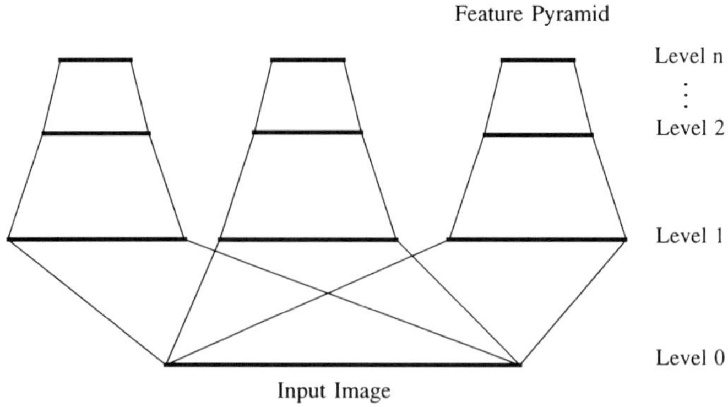


Figure 3.1: Model of hierarchical neural network

the resolution (i.e., the number of units). At a particular level $n$ ($n > 2$), a neural network classifier trained by a supervised learning method (e.g., back-propagation; Rumelhart & McClelland, 1986) uses these features to classify the image into one of $\omega$ classes. Note that we make no restrictions on the classification network. This network can be seen as an omnipotent classifier (i.e., arbitrary patterns can be assigned arbitrary classes). We show that with this architecture it is not possible to discriminate patterns, which can be discriminated at the input level (shifted versions of patterns are in the same class), at level $n$ ($n > 2$).

To simplify the notation without affecting the results, we consider the one–dimensional case. Furthermore, we can restrict our considerations to one branch of this hierarchical network. One can always select the input images so that the only relevant information is represented in one single feature. If this branch is not shift invariant, the whole system is not. It should be pointed out that the following proofs (restricted accuracy and restricted reduction functions) demonstrate the shift variance of hierarchical systems in cases where we have the freedom to choose the patterns. In cases where the number of patterns that should be processed is highly restricted, these proofs do not hold. But if we have no a priori information about the expected patterns, shift invariance cannot be guaranteed.

Let us consider a simple example of a one-dimensional hierarchical network that is shift variant. This simple example shows the basic ideas of the following proofs. Figure 3.2a shows a 3/2 pyramid where averaging is used as a reduction function. Figure 3.2b shows three input patterns $m_1, m_2, m_3$, where $m_2$ is a shifted version of $m_1$ (shift by one pixel) and $m_3$ cannot be obtained by shifting $m_1$. What we would like to have is that $m_1$ and $m_2$ be transformed into similar patterns whereas $m_3$ be transformed to a different pattern. Figure 3.2c shows the results of applying these three patterns to the 3/2 pyramid, where averaging is used as a reduction function. $m_1, m_2, m_3$ are transformed to $r_{m1}, r_{m2}$ and $r_{m3}$. One can see that $r_{m1}$ and $r_{m2}$ are considerably different from each other, whereas $r_{m3}$ is the same as $r_{m1}$. For this reason, this structure is not shift invariant, because $m_1$ cannot be distinguished from $m_3$ at the reduced level, whereas a shifted version of $m_1$ gives a considerably different result.

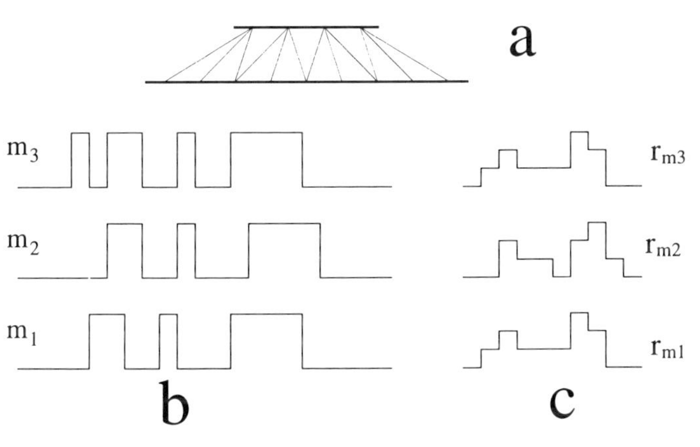


Figure 3.2: Example of shift variance for a hierarchical structure

If we can show that for any hierarchical system, there exist inputs $m_1, m_2$ and $m_3$ with $m_2 = S_d(m_1)$, for some $d$ and $m_3 \neq S_d(m_1)$, for all $d \in \mathcal{Z}$, for which the following holds: $F(m_1) = F(m_3) \neq F(m_2)$[1], then we have

[1] $S_d$ denotes a shift by a vector $d$, and $F$ denotes the reduction function of a layer of the hierarchical system.

shown that no hierarchical system is shift invariant. It should be noted that, in order to prove shift invariance, it is not sufficient to show that $F(m) = F(S_d(m))$ for all $d$. This could be easily achieved by $F(m) = 0$ for all $m$. It is also necessary that $F(m_1) \neq F(m_2)$ for all $m_1$ and $m_2$ that should be categorized differently. For example, Linhart and Dorffner (1992) noted that using special types of high order networks to obtain shift invariance reduces the discrimination ability of the following classification stage.

Unfortunately shift variance cannot be shown at the same time for arbitrary reduction functions $F$ and arbitrary accuracy of the output values. One can always choose a reduction function that encodes all possible patterns of the units in the previous level. In this case, all information available at the bottom layer is also available at higher layers and can therefore be recovered by an omnipotent classifier. For these reasons, we have to restrict either the allowed accuracy of the output values or the type of the reduction function. Both restrictions are biologically and computationally reasonable. We first prove the shift variance for a restriction of accuracy with arbitrary reduction functions and then for a restriction of the allowed reduction functions with arbitrary accuracy.

#### 3.1.1.1 Restricted Accuracy

It is sufficient to show that one level of the hierarchical system is not shift invariant, because in this case the whole system will not be shift invariant. Consider a one–dimensional array of $n$ units, where $x_i \in A; A = \{0 \ldots k-1\}$, $1 \leq i \leq n$ is the activation of the i-th unit, which is represented with an accuracy of $\log_2 k$ bits. At the next layer, we have $n/r$ ($r$ is the reduction factor) units with the same accuracy of $\log_2 k$ bits, that is, $y_i \in A; 1 \leq i \leq n/r$, where $y_i$ is the activation of the i-th unit of the reduced resolution.

The reduction function $F : A^n \mapsto A^{n/r}$ assigns to every pattern at the high resolution a new pattern at the low resolution. Note that we do not make any restrictions on the reduction function and on the connectivity.

To show that this system is not shift invariant, it is sufficient to show that we can find two patterns $m_1$ and $m_2$, where $m_2$ cannot be obtained by shifting $m_1$, so that $F(m_1) = F(m_2)$. In this case $m_1$ and $m_2$ cannot be distinguished at the lower resolution. We show this by a simple combinatorial argument:

At the reduced layer, we have $k^{n/r}$ possible patterns; at the original resolution, we have $k^n$ patterns. If we subtract those patterns that can be obtained by shifting other patterns, we have $k^{n-1}(k-1)+1$ different patterns (we assume that when a pattern is shifted, the free positions are filled with zero).[2]

In order to encode all possible configurations, $k^{n/r} \geq k^{n-1}(k-1)+1$ must hold. This yields:

$$r_s < \frac{n \log_2(k)}{(n-1)\log_2(k) + \log_2(k-1)} \leq \frac{n}{n-1} \tag{3.2}$$

for the reduction factor. Therefore, any system with $r > r_s$ (i.e., reduction by more than one unit) cannot be shift invariant, because we can always choose patterns that should be placed into different classes but are reduced to the same pattern by the resolution hierarchy. And, of course, these patterns cannot be separated, even by an omnipotent classifier, which proves that a hierarchical system with restricted accuracy is shift variant.

#### 3.1.1.2 Restricted Reduction Functions

Assume we have two successive levels with $n$ and $n/r$ units, respectively. We restrict the reduction function to compute a weighted sum of the acti-

[2]In the case of a circular shift, similar results can be obtained. The number of different patterns for a circular shift are $C(n,k) = \sum_{d|n} d|n\Phi(d)k^{n/d}$ (Rota, 1969), where $\Phi$ is the Euler function.

vations of the $m$ units of the reduction window:[3]

$$y_j = \sum_{k=jr}^{jr+m} w_{kj} x_k \tag{3.3}$$

where $x = (x_1, x_2, \ldots, x_n)^T \in \mathbb{R}^n$ is the input vector, $w_{kj}$ is the weight of the connection from unit $k$ to unit $j$, and $y_j$ is the activation of the j-th unit of the reduced layer. In the reduced layer, we have $n/r$ units. One can write equation 3.3 also in vector form:

$$y = Wx \tag{3.4}$$

where $W$ is a $(n/r) \times n$ weight matrix.

Equation 3.4 describes a linear transformation from an $n$ dimensional to an $n/r$ dimensional vector space. To show that this system is not shift invariant, it is sufficient to show that we can find two vectors $x^{(1)}$ and $x^{(2)}$ where $x^{(1)} \neq S_d(x^{(2)})$ for all $d$ and $Wx^{(1)} = Wx^{(2)}$.

The proof is straightforward: The rank of $W$ is always smaller than or equal to $n/r$; that is, $rank(W) \leq n/r$. This implies that there are no more than $n/r$ linear independent row vectors in $W$. Without restricting generality, we assume that the first $n/r$ row vectors are linearly independent. These vectors $\{r^{(1)} \ldots r^{(n/r)}\}$ form a basis of $\mathbb{R}^{n/r}$. Consider now the usual orthonormal basis of the $\mathbb{R}^n$; that is, $B = \{b^{(1)} \ldots b^{(n)}\}$, where $b^{(i)}$ is the unit vector that has a 1 at the i-th position and is zero otherwise. It is easy to see that the row vectors of $W$ can be obtained by $r^{(i)} = Wb^{(i)}, 1 \leq i \leq n$.

Because the first $n/r$ row vectors are linear independent, the row vector $r^{(i)}$ with $n/r < i \leq n$ can also be expressed as a linear combination of the first $n/r$ row vectors, that is, $r^{(i)} = \alpha_1 r^{(1)} + \cdots + \alpha_{n/r} r^{(n/r)}$. Therefore, we have for $(n/r < i \leq n)$ $Wb^{(i)} = r^{(i)} = \alpha_1 r^{(1)} + \cdots + \alpha_{n/r} r^{(n/r)} = \alpha_1 Wb^{(1)} + \cdots + \alpha_{n/r} Wb^{(n/r)} = W(\alpha_1 b^{(1)} + \cdots + \alpha_{n/r} b^{(n/r)})$. The vectors $b^{(i)}$ and $\alpha_1 b^{(1)} + \cdots + \alpha_{n/r} b^{(n/r)}$ cannot be obtained by shifting each other, which proves that this system is not shift invariant.

[3]The result holds also when we apply a nonlinear function after summation (e.g., sigmoid).

#### 3.1.1.3 Circumvention of Shift Variance

In this section, we would like to point out that the shift variance problem can be circumvented for classification tasks in a pyramid. The idea is, because we have a pyramid, we also have access to the higher resolution. If we can decide when it is necessary to access the higher resolution, we can circumvent the shift variance problem. This is very similar to the example of Landsat TM classification within a pyramid in section 2.4.3 (though the intention there was to speed up processing). One should point out that it is not always necessary to access the high resolution (e.g., in homogeneous regions), because even at the low resolution there is enough information for a decision.

Let us illustrate this fact by a simple example: Consider an object classification task, where the class of an object is determined by the number of holes. Furthermore, there might be multiple objects in the scene. The task is to classify all of these objects.

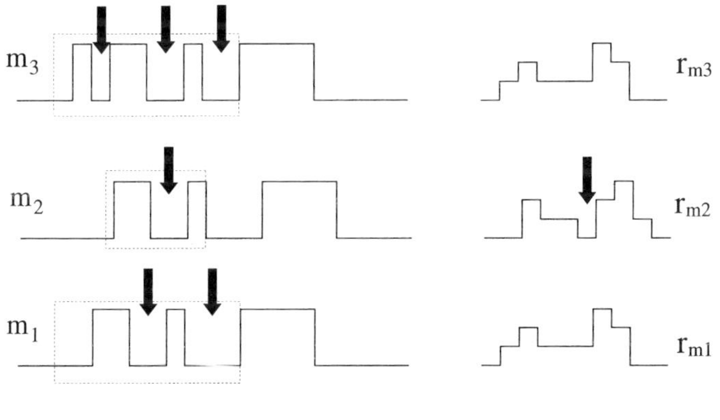


Figure 3.3: Example how to circumvent the shift variance

If the distance between the objects is larger than their size, we can easily use a pyramid to locate the objects in the image. When we have located the objects, we call an object identification network on a lower level of the pyramid. This network has to decide if it can categorize the object under consideration. If the resolution is too low, it has to call another network that examines parts of the object at a higher resolution.

Assume that we are confronted with the three patterns $m_1, m_2$, and $m_3$ shown in Figure 3.2. Figure 3.3 depicts which parts of the objects have to be examined at the high resolution (dashed box) and where the holes are detected (arrows). If a network looks at $r_{m1}$, it can already see that in the right part of the pattern there is no hole. Therefore, only the left part has to be inspected by the network below. This network will identify two holes. In the case of pattern $r_{m3}$, the network at the high resolution will identify three holes. Therefore, these two objects that have the same reduced resolution will be classified differently.

In the case of pattern $r_{m2}$, one hole can be identified at the reduced resolution, therefore, the network at the higher resolution has a smaller search space, where it finds an additional hole, so that this pattern will be assigned the same class as pattern $m_1$.

This simple example illustrates how we can circumvent the shift variance by accessing higher resolutions of the image. It also illustrates that the search space can be considerably decreased by a hierarchical representation. In order to circumvent the shift variance, we must store the whole hierarchical representation and must have access to different resolutions. The scheme we have described here is similar to the mechanism of focusing attention, which is described in more detail in section 8.2.

## 3.2 Hopfield Networks for Decimation

In this section, we introduce a modified Hopfield neural network that calculates the maximum independent vertex set of a neighborhood graph. Therefore, this network can be used as an alternative to Meers stochastic decimation algorithm. We show that the concept of the adaptive pyramid is naturally included in this framework. First, we briefly review Hopfield networks and show how they can be used for irregular decimation. We also show how to modify the network in order to get adaptive decimations. Finally, we present some experimental results.

### 3.2.1 Hopfield Networks

In 1982, Hopfield introduced a neural network model based on fully connected binary units (i.e., a unit has two states 0 or 1, and every unit is connected to every other unit except to itself in the network). The update function for a unit is:

$$s(i) = \begin{cases} 1 & \sum_j w_{ji}s(j) + I_i > U_i \\ s(i) & \sum_j w_{ji}s(j) + I_i = U_i \\ 0 & \text{otherwise} \end{cases} \tag{3.5}$$

where $s(i)$ is the state of unit $i$, $w_{ji}$ is the weight between unit $j$ and $i$, $I_i$ is the external input, and $U_i$ is a threshold value.

Hopfield proved that this system will always converge to a stable state, if the weights of the connections are symmetric (i.e., $w_{ij} = w_{ji}$) and the units update in an asynchronous manner (i.e., only one at a given time step). The proof is straightforward:

The system is governed by following energy function:

$$E = -\frac{1}{2}\sum_i \sum_j w_{ij}s(i)s(j) - \sum_k I_k s(k) + \sum_k U_k s(k) \tag{3.6}$$

This function is (for given weights, external input, and threshold) bounded, therefore, if we can prove that for every update step, we decrease the energy, we have shown that the system converges to a stable state. In order to do this, we calculate the energy difference $\Delta E$ before and after the update of a unit:

Case 1: Unit $i$ changes its state from 0 to 1.

$$\Delta E = -\frac{1}{2}\sum_j w_{ji}s(j) - \frac{1}{2}\sum_j w_{ij}s(j) - I_i + U_i \tag{3.7}$$

because $w_{ij} = w_{ji}$ we can write

$$\Delta E = -\sum_j w_{ji}s(j) - I_i + U_i \tag{3.8}$$

but in order that unit $i$ changes its state $s(i)$ from 0 to 1 $\sum_j w_{ji}s(j) + I_i > U_i$ (see equation 3.5) therefore, $\Delta E < 0$.

Case 2: Unit $i$ changes its state from 1 to 0. In this case, we can calculate $\Delta E$ in an analogous way to case 1, which yields:

$$\Delta E = \sum_j w_{ji}s(j) + I_i - U_i \tag{3.9}$$

but in order that a unit changes its state from 1 to 0 $\sum_j w_{ji}s(j) + I_i < U_i$ therefore $\Delta E < 0$.

Because the energy is bounded from below and every update step decreases the energy, we have proved the convergence of the Hopfield network.

Hopfield also gave a learning algorithm that can store a set of patterns in the network. In this case, the Hopfield network acts as an associative memory. Unfortunately the storage capacity of the network is rather limited. In McEliece, Posner, and Rodemich (1987), it was proved that for a network with $n$ binary units, the number of stored patterns that can be recovered exactly is bounded asymptotically by $\frac{n}{4 \log n}$ as $n$ approaches infinity.

In 1984, Hopfield introduced a continuous model (i.e., the state of a unit can change continuously). This model has the same characteristics as the binary model, though it is better suited for analog VLSI implementation.

Besides using a Hopfield network as an associative memory, there have been several attempts to use the network for optimization problems like the Traveling Salesman Problem (Hopfield & Tank, 1985; Kamgar-Parsi & Kamgar-Parsi, 1992) or clustering (Kamgar-Parsi, Gualtieri, Devaney, & Kamgar-Parsi, 1990). The basic idea is to formulate the optimization problem as an energy function that can be minimized by a Hopfield network. Because the update algorithm of a Hopfield network is a stochastic gradient descent search, finding a global minimum cannot be guaranteed. However, very good solutions have been reported for the continuous Hopfield model (Hopfield & Tank, 1985; Kamgar-Parsi & Kamgar-Parsi, 1992).

Our approach is similar to using the Hopfield model as an optimization procedure, however, we do not need fully connected networks, and we are not interested in the global minimum of the energy function. As shown in Theorem 1, it is sufficient to find a local minimum.

### 3.2.2 Decimation by Hopfield Networks

In the following, we show that we can replace the steps 1—3 of the algorithm for stochastic decimation (see page 35) by a modified Hopfield network that works on the neighborhood graph $G = \langle V, A \rangle$.[4] Moreover, we show that the formulation as a Hopfield network is more general than stochastic decimation, and it naturally includes the concept of the adaptive pyramid.

Let us introduce the notion of a survival state of a cell:

**Definition 3** *The survival state of a cell $p \in V$ is a function*

$$s : V \mapsto \{0, 1\} \textit{ with } s(p) = \begin{cases} 1 & \textit{if cell } p \textit{ survives} \\ 0 & \textit{otherwise} \end{cases}$$

Let us further introduce the following energy function:

$$E = \sum_{\langle i,j \rangle \in A} s(i)s(j) - \sum_{k \in V} s(k) \tag{3.10}$$

Now the following theorem holds:

**Theorem 1** *The energy function $E$ from equation 3.10 obtains a local minimum, $E_{min}$, if and only if the assignment of surviving and non-surviving cells, $s(p)$, is a valid decimation (i.e., satisfying the rules 1 and 2 of definition 2) or, equivalently, forms a maximum independent vertex set of $G$.*

[4]We skip the sub-indices for level $i$ because the algorithm works only on one level.

*Proof.*
(a) $E_{min}$ is a local minimum of $E \Rightarrow s(p)$ is a valid decimation.
Assume $E = E_{min}$ is a local minimum but $\{p \in V | s(p) = 1\}$ is not a valid decimation, then at least one of the rules 1 or 2 must be violated.

Case a.1: Rule 1 does not hold:

$\Rightarrow \exists\ p, q \in V$ such that $\langle p, q \rangle \in A$ and (by definition 3) $s(p) = s(q) = 1$. Changing $s(p)$ to 0 can affect only those terms in equation 3.10 where $s(p)$ occurs. We can rewrite $E$ as

$$E = \sum_{\langle p,n \rangle \in A} s(p)s(n) + \sum_{\langle n,p \rangle \in A} s(n)s(p) + \sum_{\substack{\langle i,j \rangle \in A \\ i,j \neq p}} s(i)s(j) - s(p) - \sum_{k \neq p \in V} s(k) \tag{3.11}$$

The fourth term increases the energy by 1 if we change $s(p)$ to 0. But the first and second term decrease the energy at least by 2 because $s(p)s(q)$ changes from 1 to 0 in both terms. Following inequality holds: $E(s(p) = 0, s(q) = 1) \leq E_{min} + 1 - 2 = E_{min} - 1 < E_{min}$ and this is a contradiction that $E_{min}$ is a local minimum.

Case a.2: Rule 2 does not hold:

$\Rightarrow \exists p \in V$ such that $s(p) = 0$ and $\forall q \in \Gamma(p) :\ s(q) = 0$. Again changing $s(p)$ to one yields $E(s(p) = 1) = E_{min} - 1 < E_{min}$ and this is a contradiction to the fact that $E_{min}$ is a local minimum.

(b) $\{p | s(p) = 1\}$ defines valid decimation $\Rightarrow E = E_{min}$ is a local minimum

We have to show that by changing only one state of a cell $r \in V$, we get a higher energy value.

b.1: $s(r) = 0 \Rightarrow \exists q \in \Gamma(r) :\ s(q) = 1 \Rightarrow E(s(r) = 1) = E_{min} - 1 + 2 = E_{min} + 1 > E_{min}$

b.2: $s(r) = 1 \Rightarrow \forall q \in \Gamma(p) :\ s(q) = 0 \Rightarrow E(s(r) = 0) = E_{min} + 1 > E_{min}$

From b.1 and b.2 we conclude that $E_{min}$ is a local minimum of $E$. qed.

Given the energy function in equation 3.10 we can now define a Hopfield network operating on the neighborhood graph that minimizes this energy function. If we set in equation 3.6

$$w_{ij} = \begin{cases} -2 & \langle i,j\rangle \in A \\ 0 & \text{otherwise} \end{cases} \tag{3.12}$$

and $I_i = 1$ and $U_i = 0$ for all $i$, we get equation 3.10. The resulting Hopfield network operates on the neighborhood graph $G$ and computes valid decimations (according to definition 2). The update procedure for the cells is as follows:

$$s(p) = \begin{cases} 1 & \text{if } 1 - 2\sum_{\langle q,p\rangle} s(q) > 0 \\ 0 & \text{otherwise} \end{cases} \tag{3.13}$$

The initial states of the cells can be chosen at random.

We can extend our results of Theorem 1 to include the weights and the external input explicitly. Let us define following energy function:

$$E = -\frac{1}{2}\sum_{\langle i,j\rangle \in A} w_{ij}s(i)s(j) - \sum_{k \in V} I_k s(k) \tag{3.14}$$

Now we can prove the following theorem:

**Theorem 2** *The energy function $E$ from equation 3.14 obtains a local minimum, $E_{min}$, with $I_k > 0\ \forall k \in V, w_{ij} = w_{ji} < 0$ and $I_k < |w_{ij}|\ \forall k \in V, \langle i,j\rangle \in A$ if and only if the assignment of surviving and non-surviving cells, $s(p)$, is a valid decimation (i.e., satisfying the rules 1 and 2 of definition 2) or, equivalently, forms a maximum independent vertex set of $G$.*

The proof is nearly identical to the proof of Theorem 1, therefore we omit it here; it can be found in Bertin and Bischof (1993). This result states essentially that all networks that satisfy the conditions on the weights and external input of Theorem 2 form an equivalence class. That is, as long as the conditions of the theorem are satisfied, the network will end up in the same final state (the initialization and update order must also be identical).

The update function of the cells with the more general form of the theorem is:

$$s(p) = \begin{cases} 1 & \text{if } I_p + \sum\limits_{\langle q,p\rangle} w_{qp}s(q) > 0 \\ 0 & \text{otherwise} \end{cases} \tag{3.15}$$

Having established the relationship between Hopfield networks and selection of survivors, we can now apply the theory available for Hopfield networks. For example, the convergence time is of interest because in the algorithm of Meer the convergence takes $\mathcal{O}(|V|)$ steps in the worst case (when the random number generator produces a ramp function on every iteration). In Shrivastava, Dasgupta, and Reddy, (1992), it was proven that a Hopfield network with negative weights (if the weights are 0, the connection is not present) and a sequential update algorithm converges in the worst case in $2|V|$ steps.

But one should note that by a parallel update scheme and a connectivity graph that is far away from full connectivity, one could design a parallel algorithm that converges much faster, because all units that are not connected can be updated in parallel without altering the convergence properties.

According to the theorems, we can produce all possible maximum independent vertex sets of a graph with a Hopfield network. Which set we produce depends on the (random) initial state and the (random) order of update of the cells. Because the Hopfield network performs a stochastic gradient descent search on the energy function $E$, the result is also stochastic. Therefore, the resulting pyramid structure will also be stochastic as in the case of Meer's algorithm. There is no use in performing more sophisticated procedures like simulated annealing or mean field theory annealing (Bilbro et al., 1989; Geman & Geman, 1984; Kirkpatrick, Gelatt, & Vecchi, 1988; Peterson & Anderson, 1987; Soukoulis, Levin, & Grest, 1983), because the global minimum of the energy function is unique (up to symmetries) and corresponds to the maximum number of surviving cells. In this case, the pyramid would no longer be stochastic. For example, in the case of a four–connected pixel grid, we would get the $3 \times 3/2$ pyramid, which is regular and therefore shift variant.

However, there exist other types of minimization procedures that can be used, for example, a simple greedy algorithm that calculates for every cell the energy difference that is gained by changing the state of the cell and then selects the cell with the largest energy difference for update. This algorithm has the disadvantage that it is not local.

The other advantage of the formulation as a Hopfield network is that already chips for simulating Hopfield networks are available, so it is easy to implement the selection of survivors in hardware.

The essential result of this section is that we can specify an energy function for selecting the survivors. This energy function might be useful for certain proofs about irregular pyramids because it is mathematically easier to handle than the algorithm of Meer.

### 3.2.3 Adaptive Pyramid

In this section, we demonstrate how to influence the decimation process by a proper setting of weights of the Hopfield network. If the weights depend on the content of the image, the resulting pyramid structure will be adapted to the image. Therefore, it is called adaptive pyramid.

The energy function 3.14 has the weights $w_{ij}$ and the external input $I_k$ as parameters that can be changed in order to influence the decimation. The weights $w_{ij}$ between two cells express the constraints on the mutual states of these cells. If $w_{ij}$ is negative these two cells should not be both on; on the other hand, a positive weight forces the two cells to be both on. One should note as long as $w_{ij} > -I_j$, the behavior of the network is not changed (as can be seen from Theorem 2). The external input $I_i$ can force a single cell to survive or not to survive (if $I_i > \sum_{j \in \Gamma(i)} |w_{ji}|$ cell $i$ survives, if $I_i < 0$ [and the weights are negative], cell $i$ will not survive).

We can now use these observations to build adaptive pyramids (Jolion & Montanvert, 1992; Montanvert & Bertolino, 1992). To determine the connected components of a binary image, we can set $I_k = 1$ for all $k \in V$, and the weights according to:

$$w_{ij} = \begin{cases} -2 & i \text{ and } j \text{ are neighbors and have the same grey value} \\ 0 & \text{otherwise} \end{cases} \tag{3.16}$$

or expressed differently $w_{ij} = -2(1 - d_{ij})$, where $d_{ij} = |g_i - g_j|$, $g_i \in \{0, 1\}$ are the grey values of the pixels. In this case, the survivors and non-survivors are computed only within a homogeneous region (with identical pixel values). This procedure is equivalent to Montanvert's algorithm (Montanvert & Bertolino, 1992), which applied the stochastic decimation only within homogeneous regions. However, it should be noted that we have only changed the weights and do not alter the algorithm nor the graph, as is the case for the algorithm of Montanvert.

We can now generalize this scheme to grey level images. We set the weights of the corresponding Hopfield network according to the difference in grey value of the pixels, that is:

$$w_{ij} = -f(d_{ij}) \text{ and } d_{ij} = |g_i - g_j| \tag{3.17}$$

A generalization of function 3.16 is

$$w_{ij} = -f(d_{ij}) = \begin{cases} 2 & d_{ij} < \Theta \\ 0 & \text{otherwise} \end{cases} \tag{3.18}$$

a threshold function. A little bit more general function is:

$$w_{ij} = -f(d_{ij}) = \begin{cases} 2 & d_{ij} < \Theta_1 \\ \frac{2(\Theta_2 - d_{ij})}{(\Theta_2 - \Theta_1)} & \Theta_1 \le d_{ij} \le \Theta_2 \\ 0 & d_{ij} > \Theta_2 \end{cases} \tag{3.19}$$

In this case, we treat the regions as homogeneous if $d_{ij} < \Theta_1$ and different if $d_{ij} > \Theta_2$.

In choosing the function $f$ for setting the weights, one must be careful that $|w_{ij}|$ is not too small, because when $I_j > \sum_i |w_{ij}|$ the cell will always survive, but this should happen only in those cases where there is no similar cell in the neighborhood.

It is interesting to note that the external input $I_j$ can control the number of surviving cells (and therefore the regions found). If $I_j$ is large many cells will survive; if it is small, fewer cells will survive, where for positive $I_j$ the limit is a maximum independent vertex set.

With the adaptive pyramid and the setting of weights we are influencing the decimation ratio, that is, allowing more cells to survive. It is important that we do not alter the property that a non-survivor must be a neighbor of a survivor, because this would make the algorithm nonlocal, which implies that a parallel implementation would be very costly. The modifications we have proposed leave the decimations in homogeneous regions unchanged, but in non-homogeneous regions, more cells are allowed to survive, so that a non-surviving cell has the possibility to select a similar cell as father.

This is also the basis for the invariance property of adaptive pyramids. Because the survivors depend on the content of the image, if we shift the image, then the survivors in non-homogeneous regions will also be shifted, whereas in homogeneous regions, it does not matter which cell survives.

The algorithm for the adaptive pyramid controlled by a Hopfield network is as follows:

1. Calculate the neighborhood graph of the base level.
2. Initialize the states of the cells at random and compute the weights according to the function $f$.
3. Update the states of the cells in an asynchronous manner according to the update function 3.15 until convergence.
4. Every non-surviving cell ($s(i) = 0$) selects the father with the most similar grey value.
5. Calculate the new neighborhood graph of the next level (according to step 5 of the algorithm of the stochastic pyramid).
6. Repeat steps 2-6 until the graph does not change any more (i.e., all cells are survivors).

Compared to the algorithms of Jolion and Montanvert (Jolion & Montanvert, 1992; Montanvert & Bertolino, 1992), the formulation with the Hopfield network has the advantage that we do not need to calculate the similarity graph and can operate on the neighborhood graph, which is symmetric. Therefore, we say that the decimation by the Hopfield network includes naturally the concept of the adaptive pyramid. In the next section, we demonstrate our theoretical results by several experiments.

### 3.2.4 Experiments

We first demonstrate the validity of Theorems 1 and 2. For these simulations, we have used a small image with $30 \times 30$ pixels. We show the decimation only for the base level of the pyramid. Figure 3.4 depicts a decimation for a 4-connected (left) and an 8-connected (right) neighborhood graph. The decimations are all valid, as is expected from Theorem 1. In the case of 4-neighborhood, one can see that regular patterns, which are the dominant structure, are broken by line-like structures. This resembles somehow the structure of the human retina, where one can see similar patterns (Kropatsch, 1992). The network converges on the average in 8 asynchronous update steps. The maximum number during our simulations (100 runs) were 12 update steps.

Figure 3.5 shows the decimation for a binary image with 8-neighborhood where we have used a line from left to right at columns 13–15. The weights are set according to function 3.16. The Hopfield network treats the components of the image independently, that is, the line and the background are decimated independently of one another.

Within the framework of irregular pyramids, we are not constrained to start from a regular pixel grid; this advantage is used in the following. We first apply an adaptive Voronoi tessellation to the image. And then we construct on this already irregular structure (we use the Delaunay triangulation, which is the dual of the Voronoi tessellation) an irregular pyramid. The advantage of this approach is that the Voronoi tessellation and the Delaunay graph offer a reduced description of the image and the neighborhood relation between regions, which is already adapted to the image

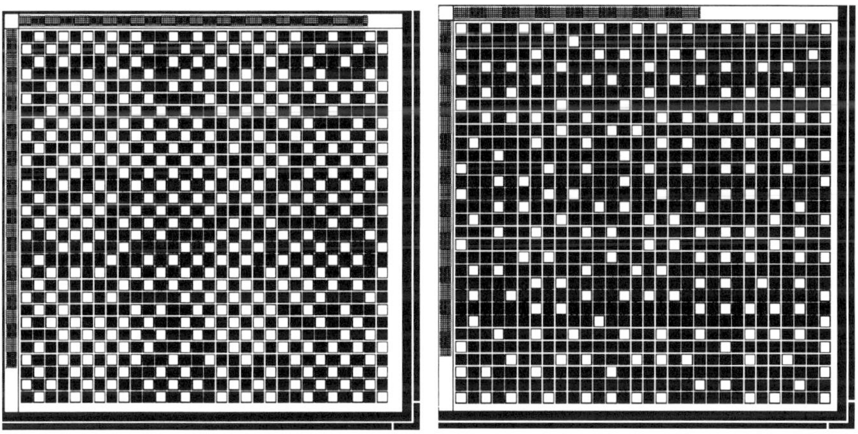

(a) (b)

Figure 3.4: Decimations of graphs with 4-neighborhood (a) and 8-neighborhood (b)

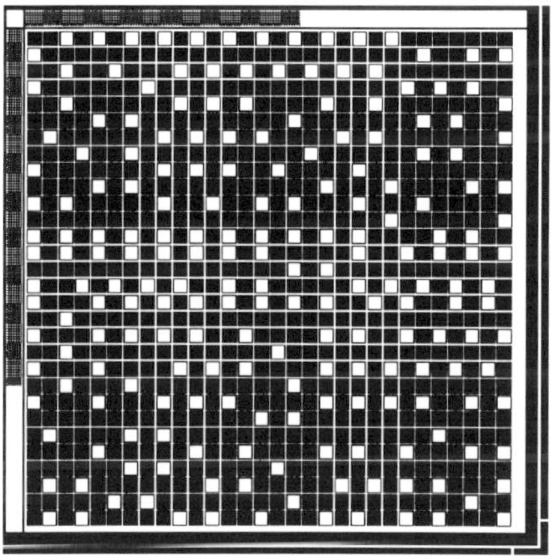

Figure 3.5: Decimation of a binary image

content. This results in pyramids with less levels, compared to starting with the original image. The Delaunay graph has the further advantage that it is planar (this is not the case for the 8–connected graph), and that the planarity is preserved by the pyramid construction (as was proved in Kropatsch & Montanvert (1992)). We first describe the algorithm for the adaptive Voronoi tessellation and then the construction of irregular pyramids on top of it.

#### 3.2.4.1 Adaptive Voronoi Tessellation of an Image

The algorithm for computing the Voronoi diagram and Delaunay graph is a generalization of the split and merge algorithm on quadtree regular (rigid) structures, extended to Voronoi irregular (non-rigid) structures and is due to Bertin (Bertin, Parazza, & Chassery, 1993). This algorithm proceeds in two steps. The first step is splitting: Polygons are added in the support of the image until convergence. The second step is the merge step: Some polygons are deleted. A polygon is added or deleted according to the following definitions:

**Definition 4** *A region enclosed by a polygon $P$ is said to be non-homogeneous if and only if the variance in the region is larger than a given threshold.*

**Definition 5** *A polygon $P$ is said to be useless if and only if all the neighbors of $P$ have almost equal grey level means.*

More precisely, the algorithm proceeds in following major steps (for more details, see Bertin et al. (1993)):

1. Assign five points in the image (one point in the center and four on the corners).
2. Compute Voronoi diagram and Delaunay graph.
3. Compute mean grey value, standard deviation, and surface of each polygon.

4. For each polygon, split if it is not homogeneous.

5. Repeat 2–4 until convergence (i.e., all polygons are homogeneous).

6. Merge: suppression of useless polygons.

The proof of convergence is straightforward (in the worst case, all the polygons have one pixel in their interior that gives a variance equal to zero).

An illustration of this algorithm is given in Figure 3.6, where the size of the image is $256 \times 256$ with 256 grey values.

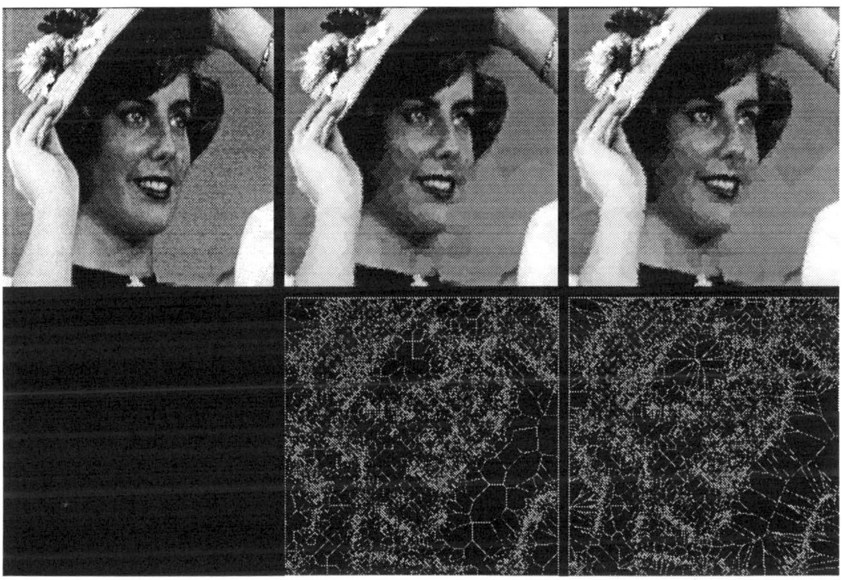

Figure 3.6: Original image (left), split result (10,585 polygons) and the graph (middle), merge result (9,033 polygons) and the graph (right).

#### 3.2.4.2 Image Segmentation With Voronoi Pyramids

The algorithm proceeds in the following major steps.

1. Compute an adaptive Voronoi partition of the image with the split and merge algorithm.

2. Compute the pyramidal process with the adaptive decimation with Hopfield neural networks.

In Figure 3.7, we have applied the decimation with the Hopfield network to the Delaunay graph of Figure 3.6 and built a stochastic pyramid on top of it. In these experiments the convergence time of the Hopfield network was in the average about five steps.

Figure 3.8 shows the segmentation of "Lena" with our method. The images are of size $256 \times 256$. The total time to process such an image is around 2 minutes on a Sun 4 workstation.

Figure 3.9 shows a comparison between results of our method with the results obtained by a stochastic pyramid. The parameters for Figure 3.9a are: $\Theta_1 = 27$ and $\Theta_2 = 57$. We have used a threshold of ($\Theta = 27$) to compute the similarity graph of the stochastic pyramid (Figure 3.9b) The results are at a first glance rather similar (see Figure 3.9c,d for a comparison of the end results). However, the number of levels and the number of regions is higher than for the pyramids using the Delaunay graph as the base level. For the stochastic pyramid, we need 11 levels ending up with 17 regions, whereas we need for our method only 9 levels ending up with 9 regions.

These experimental results demonstrate the validity of our approach and show that the decimation by Hopfield networks is an alternative to the algorithm of Peter Meer.

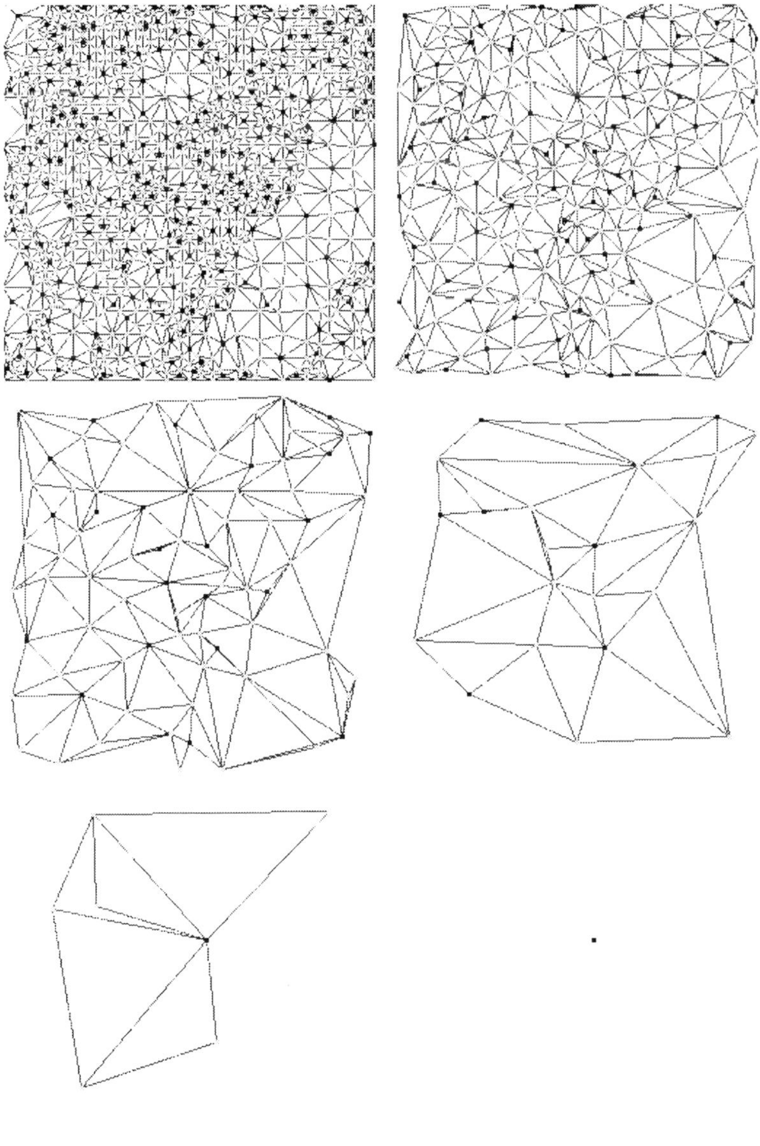

Figure 3.7: Irregular pyramid, starting with a Delaunay graph (Black dots indicate survivors)

Figure 3.8: Image "Lena," final result of the segmentation. Starting with 9,033 Voronoi polygons, 50 regions remain.

## 3.3 Summary

In this chapter, we have shown that the structures of image pyramids and neural networks are similar. In particular, for any image pyramid an equivalent neural network topology can be built. We have used this knowledge throughout this chapter. First, we have shown that shift variance is also a problem for hierarchical neural networks. We have proven this in a general way.

In the next section, we have analyzed irregular pyramids, because they offer a way to avoid the shift variance problem. We have introduced an energy function and proved that any local minimum of this energy function corresponds to a maximum independent vertex set. We have shown how this function can be minimized by Hopfield neural networks. We have also demonstrated that the Hopfield network formulation includes naturally the concept of the adaptive pyramid. Several examples have demonstrated this approach and how it might be used for image segmentation.

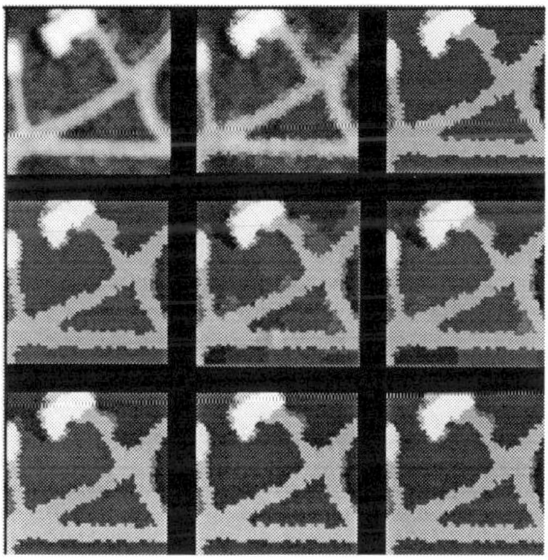

(a) Adaptive segmentation with function (3.19). From left to right and top to bottom: Original image. Voronoi diagram: 1043 polygons. Level 8: 12 regions. Final result (level 9): 9 regions. Level 3: 129 regions. Level 4: 123 regions. Level 5: 59 regions. Level 6: 34 regions. Level 7: 19 regions.

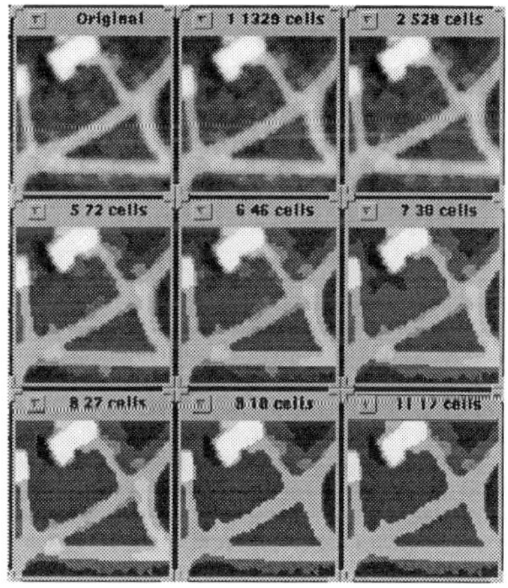


(b) Stochastic pyramid segmentation. From left to right and top to bottom: Level 0: Original image 4096 pixels. Level 1: 1329 regions. Level 2: 528 regions. Level 5: 72 regions. Level 6: 46 regions. Level 7: 30 regions. Level 8: 19 regions. Final result (level 11): 17 regions.

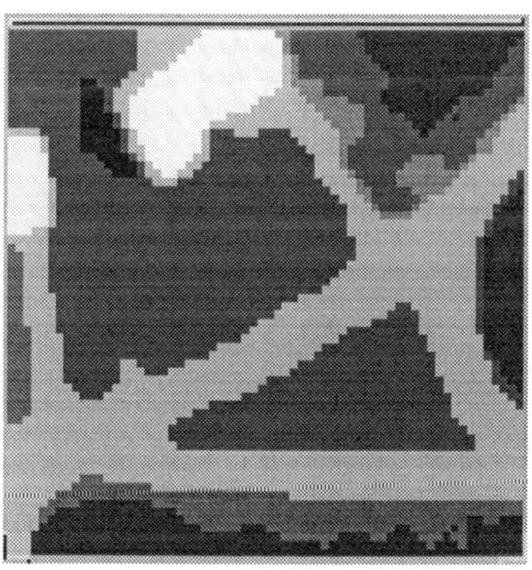

(c) Stochastic pyramid

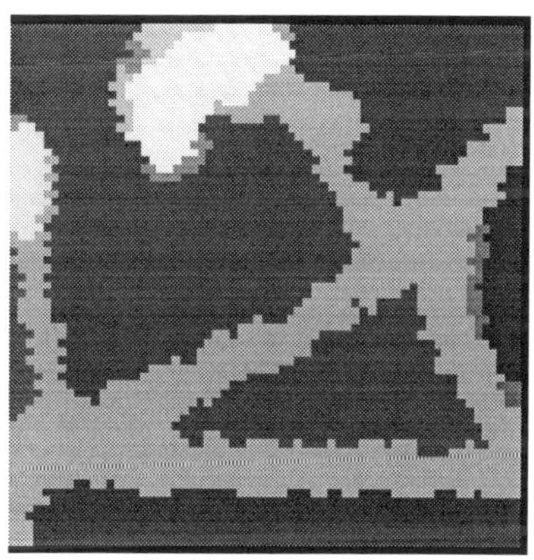

(d) Adaptive pyramid

Figure 3.9: Comparison of segmentation results

# Chapter 4

# Contents and Representation

A fundamental question for all vision systems is which information should be stored in the system and how should it be represented. The answer to the first question is mainly determined by the task to solve, whereas the answer to the second question depends on the algorithms used to solve the task at hand. The representational issue is also a fundamental question for neural networks (Fodor and Pylyshyn, 1988; Hancock, 1988; Hinton et al., 1986; Pollack, 1990; Smolensky, 1990). How should we represent information in units? This question is extremely important because units are restricted in the functions they can perform, and we cannot use complex data structures as is the case for conventional algorithms. We have to represent any type of information as a pattern of activity among the units.

Neural network researchers have often ignored this question, assuming they have a vector (or list) of values they should process. Of course, any type of information can be coded in the form of some suitable vectors, however, often valuable information is hidden by this process of encoding and cannot be easily accessed. This information has then to be recovered by some learning algorithm, which may take a very long time.

One example is using images (or parts of images) as input to a neural network. One valuable sort of information are geometric relations of the pixels (i.e., which pixel is neighbor of another pixel), but this information is completely hidden by a vector encoding scheme. It can only be recovered by a learning algorithm by noticing the correlations of adjacent pixels. But in order to recover this obvious information, thousands of training examples and a lot of training time is necessary.

Image pyramids do not have this problem, because due to the structure of the pyramid, every cell "knows" its position and neighbors. In an image pyramid, an image is represented in a retinotopic way. This means that the topology of the bottom level (retina) is to a large extent preserved in higher levels. This is also common in biological visual systems. Neural network models that have used this kind of representation are used with great success for image pattern recognition (Fukushima et al., 1983; Le Cun et al., 1990b). It is not surprising that these structures are very similar to image pyramids.

We have seen in the previous chapter that we can build network topologies that are equivalent to image pyramids. Therefore, we can also use the retinotopic representation in neural networks. However, we have seen in section 2.2.2 that we can store numeric and symbolic information in a pyramid, whereas neural networks store only numeric information in the units. Therefore, cells of a numeric pyramid (especially grey level pyramids) are from the viewpoint of contents rather similar to neural networks. We use this similarity further in chapters 5 and 6. The case of symbolic information, however, needs further investigations.

## 4.1 Symbolic Content of a Cell

Let $\Sigma = \{\sigma_1, \sigma_2, \ldots \sigma_N\}$ be a finite set of symbols and $R \subseteq \Sigma \times \Sigma$ a binary relation between these symbols. A cell in a symbolic pyramid stores symbols and/or relations between these symbols. Our main concern in this section is how we can represent this information by a neural network. We are therefore confronted with a representation problem.

Because we have a finite number of symbols (and therefore also of relations), we can code a symbol by the activation of a single unit, for example, $\sigma_1 = 0, \sigma_2 = 0.1$, and so on. This representation is very "unnatural" for neural networks. When we want to process these symbols, we have to include a complex coding/decoding mechanism into the units. Therefore, the processing performed by a unit would not be "simple," and this is an essential characteristic of neural networks we do not want to give up. The only way to avoid this problem is to use several units to represent a symbol. The consequence is that we have to replace one cell of a symbolic pyramid by a set of units, so we give up the one–to–one correspondence between cells and units (as is the case for grey level pyramids). We have two possibilities for the kind of representation (see section 2.1.3):

1. Local representation
2. Distributed representation

## 4.2 Local Representation

If we have $N$ symbols, we need $N$ units to represent these symbols locally. Every symbol has a designated unit that becomes active when the symbol is present. If we have to represent binary relations among the symbols, we need $\mathcal{O}(N^2)$ units to represent all possible relations. When only one symmetric relation at a time needs to be represented, $N$ units are sufficient (i.e., if both units $\sigma_1$ and $\sigma_2$ are active, the relation $\sigma_1 R \sigma_2 = \sigma_2 R \sigma_1$ is represented). When we store more than one relation at a time, say $\sigma_1 R \sigma_2$ and $\sigma_3 R \sigma_4$, we can no longer discriminate the relations (e.g., $\sigma_1 R \sigma_4$ and $\sigma_2 R \sigma_3$ would give the same representation). This is a classic instance of the so–called binding problem (Fodor & Pylyshyn, 1988). We need therefore a designated unit for each relation $\sigma_i R \sigma_j$ that is active when the relation is present. The problem even becomes more severe when several different relations or multi-valued (n-ary) relations need to be represented.

## 4.3 Distributed Representation

We have seen that we run into a problem of combinatorial explosion with a local representation. Distributed representations are more economical in terms of units and are also more robust than local representations (Hinton et al., 1986). The general idea is that more than one unit is active when representing a symbol (relation), and a single unit participates in representing more than one symbol (relation). Many possibilities for distributed representations have been proposed (Ballard, 1987; Hancock, 1988; Hinton, 1990; Hinton et al., 1986; Pollack, 1990; Smolensky, 1990). Many of them are suited for representing numeric information (Ballard, 1987; Hancock, 1988); others have been proposed for symbolic information (Hinton, 1990; Pollack, 1990; Smolensky, 1990).

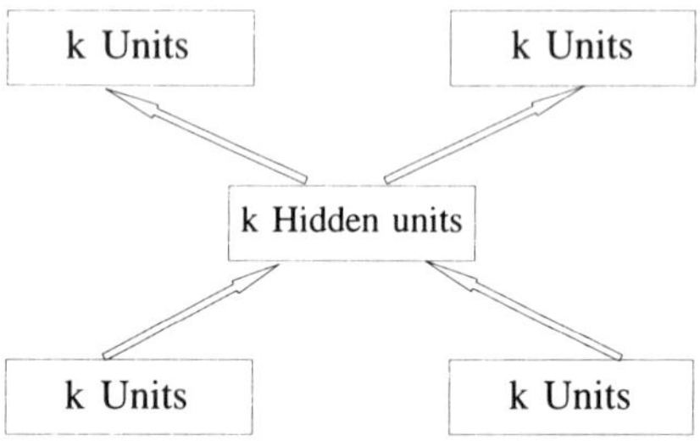


Figure 4.1: Auto-associative network of Pollack

Pollack (1990) introduced the concept of "Recursive Distributed Representations." He showed how to represent binary trees and arbitrary list structures in this framework. The general idea is very simple: The basic network is an auto-associative network[1] like the one in Figure 4.1. This network is trained to reproduce the input at the output units through a narrow channel of hidden units (in this case half the number of hidden units than input units). Consider how to represent the binary tree in Figure 4.2 in that network.

Assume that the symbols at leave of the tree $\{a, b, c, d\}$ are coded in a local way with ($K \geq 4$ units). First we learn to auto-associate two leaves of the tree, that is, $\{(a, b), (c, d), (d, c)\}$. If the network has learned this task, we consider the activations of the hidden units as a distributed code

[1]We analyze the behavior of auto-associative networks in more details in chapter 6.

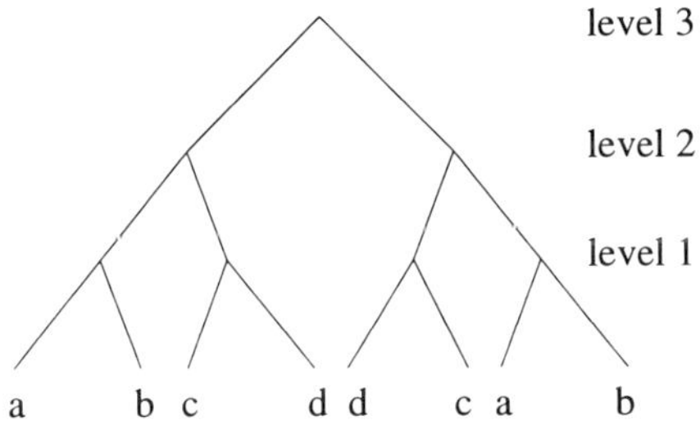


Figure 4.2: A binary tree to represent in the Pollack network

for the nodes of level 1. Now we add these codes to the training set (i.e., $H(a,b), H(c,d), H(d,c)$, where $H(a,b)$ are the activations of the hidden units when $a$ and $b$ are represented at the input), and continue training until the network has learned the task. Then we proceed in the same way with the next level and so on, until we have reached the root of the tree. If the network is able to learn this task, we have essentially coded the whole tree structure with a code of fixed length in a distributed form. We can now decode the tree: Assume $H_R$ is the code of the root of the tree. When we set the hidden units of the network in Figure 4.1 to $H_R$ and update the output units, we will get at the output units the code for the nodes of level $H_{R-1}$. We can now use these codes as activations for the hidden units, and we will get the next lower level, until we have reached the leaves of the tree.

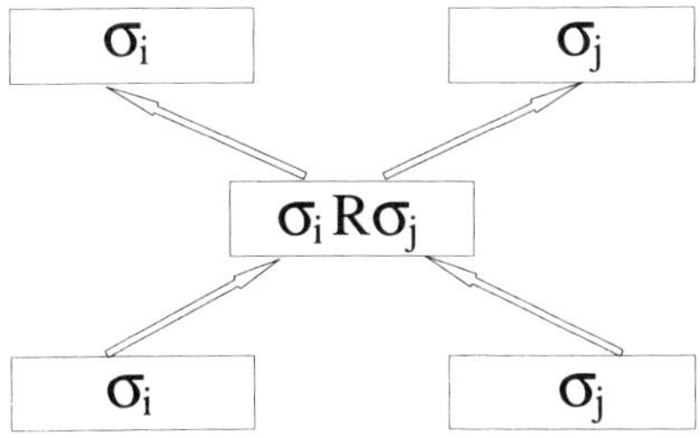


Figure 4.3: Network for distributed representation of relations

For our purpose, we can use this scheme to represent relations with only $\mathcal{O}(N)$ units ($N$ is the number of symbols to represent). We need only the first step of the Pollack network, and we do not need to apply it recursively. Assume that we have local representation of the individual symbols. We can now represent the relation $\sigma_i R \sigma_j$ with $N$ units like in Figure 4.3. Assuming a binary code develops in the hidden units (which is not quite true as we will see in the experiment later), we can represent all relations with this scheme, because $N \neq 3$ $2^N \geq N^2$. In a recent paper (Phatak, Choi, & Koren, 1993), it has been proven that for the encoder problem with $n$ input units and $n$ input patterns 2 hidden units are sufficient for arbitrary $n$. Therefore, we can safely assume that we can always represent the relations with the proposed auto-associative network.

With this scheme, we can convert the distributed code back into a local one. Moreover, we can use constraints on the weights for special kinds of relations. For example, if the weights from the first set of input units are identical to the second set of input units, the representation of the relation is symmetric, that is, $\sigma_i R \sigma_j = \sigma_j R \sigma_i$.

## 4.4 Experiment

In order to see which code develops at the hidden units, we have performed the following experiment. We use 4 symbols $(a, b, c, d)$ represented locally at the input and output units. The network has 8 input units, 4 hidden units and 8 output units. Because we want to represent a symmetric relation, we have constrained the weights from input units $1-4$ to the hidden units to be identical to the weights of input units $5-8$ to the hidden units (e.g., $w_{11} = w_{51}$). The network is trained with back–propagation to auto-associate the input with the output. This takes about 100 presentations of training patterns until the network converges. In Table 4.1, we show the activations of the hidden units for the inputs. One can see that the hidden units have developed a distributed code that is nearly binary. In Table 4.2, we show the Euclidean distance between the code vectors. One can see that the distance between codes that have one symbol in common is usually smaller than for those that do not. In section 5.2, we discuss how this code might be used to process symbolic information in a pyramid.

| | | | | |
|---|---|---|---|---|
| aa | 1 | 0 | 0 | 1 |
| ba | 1 | 0.15 | 0 | 0.2 |
| ca | 0 | 0 | 0 | 1 |
| da | 1 | 0 | 1 | 1 |
| bb | 1 | 1 | 0 | 0 |
| cb | 0 | 1 | 0 | 0 |
| db | 1 | 1 | 1 | 0 |
| cc | 0 | 1 | 0.75 | 1 |
| dc | 0 | 0 | 1 | 0.54 |
| dd | 1 | 0 | 1 | 0 |

Table 4.1: Code Developed by Auto-associative Network

| $d^2$ | aa | ba | ca | da | bb | cb | db | cc | dc | dd |
|---|---|---|---|---|---|---|---|---|---|---|
| aa | 0 | 0.66 | 1 | 1 | 2 | 3 | 3 | 2.06 | 2.21 | 2 |
| ba | 0.66 | 0 | 2.36 | 1.66 | 0.89 | 1.89 | 1.89 | 2.93 | 2.18 | 1.06 |
| ca | 1 | 2.36 | 0 | 2 | 3 | 2 | 4 | 1.56 | 1.21 | 3 |
| da | 1 | 1.66 | 2 | 0 | 3 | 4 | 2 | 2.06 | 1.21 | 1 |
| bb | 2 | 0.89 | 3 | 3 | 0 | 1 | 1 | 2.56 | 3.21 | 2 |
| cb | 3 | 1.89 | 2 | 4 | 1 | 0 | 2 | 1.56 | 2.29 | 3 |
| db | 3 | 1.89 | 4 | 2 | 1 | 2 | 0 | 2.06 | 2.29 | 1 |
| cc | 2.06 | 2.93 | 1.56 | 2.06 | 2.56 | 1.56 | 2.06 | 0 | 1.27 | 3.06 |
| dc | 2.21 | 2.18 | 1.21 | 1.21 | 3.21 | 2.29 | 2.29 | 1.27 | 0 | 1.21 |
| dd | 2 | 1.06 | 3 | 1 | 2 | 3 | 1 | 3.06 | 1.21 | 0 |

Table 4.2: Distance of Code Vectors

## 4.5 Summary

In this chapter, we have analyzed the content of pyramidal cells. We have seen that the representation of information in pyramids is done in a retinotopic fashion. For the case of numeric information, there is not much difference between pyramids and neural networks. For the case of symbolic information, we had to give up the one–to–one correspondence between a cell and a unit. A pyramidal cell is replaced by a small neural network in order to represent symbolic information. We have discussed local and distributed representations. In the case of distributed representations, we have developed a network that can learn how to represent relations between symbols in a distributed manner.

# Chapter 5

# Processing

In this chapter, we investigate the processing performed by pyramid cells and how it might be transferred to hierarchical structured neural networks. We analyze the neural network solution and see if it can provide anything new for pyramids. We restrict ourself to bottom-up processing, however, similar results can also be obtained for top-down processing. Because there is no use in performing symbolic processing on numeric information and vice versa, we consider processing of numeric and symbolic information separately.

## 5.1 Numeric Processing

In section 2.2.3.1 we have seen that *linear filters*, *nonlinear filters*, and *morphological filters* are the most commonly used reduction functions for pyramids. Let us discuss how they can be implemented by neural networks.

Because linear filters can be expressed as convolutions, and this is the same as the update function of most neural networks (weighted sum), we can easily perform this operation with neural networks. Let us note two differences: In pyramids usually the same kernel is used for all cells, whereas in neural networks every unit has different weights. But there are some models (e.g., weight sharing; Le Cun et al., 1990b) that force units to develop identical weights. Another difference is the nonlinearity (typically a sigmoid function), which is used by most neural networks after the weighted sum. Let us investigate the effect of such a function for pyramids.

### 5.1.1 Sigmoid Pyramid

In image processing terms, the application of a sigmoid function corresponds to a nonlinear contrast enhancement operation. Let us call pyramids that use a convolution with a sigmoid contrast stretch as a reduction function *sigmoid pyramids*.

Because we want to process grey-values in the range $[0 \ldots 255]$, we have to adapt the sigmoid $f(x) = \frac{1}{1+e^{-x}}$ to that range. In particular, we want $f(0) \approx 0$ and $f(255) \approx 255$. A little bit of calculation yields the following formula:

$$f(x) = \frac{255}{1 + e^{-\frac{14x}{255}+7}} \tag{5.1}$$

In order to control the steepness of the function, an additional parameter $\beta$ is introduced:

$$f(x) = \frac{255a(\beta)}{1 + e^{\beta(-\frac{14x}{255}+7)}} - b(\beta) \tag{5.2}$$

$a(\beta)$ and $b(\beta)$ are chosen such that $f(0) \approx 0$ and $f(255) \approx 255$. This yields:

$$a(\beta) = \frac{1 + e^{7\beta}}{e^{7\beta} - 1} \tag{5.3}$$

$$b(\beta) = \frac{255}{e^{7\beta} - 1} \tag{5.4}$$

For $0 < \beta < 1$ $f(x)$ is flat, and for $\beta > 1$ $f(x)$ gets steeper. In the limit $\beta \to 0$, $f(x)$ approaches a linear function and for $\beta \to \infty$, $f(x)$ approaches a step function. Figure 5.1 shows $f(x)$ for various values of $\beta$.

Let us describe qualitatively the effect of applying $f(x)$ to an image. Consider an image with an equally distributed histogram. The application of $f(x)$ will increase the histogram at the borders and decrease it in the middle. The effect on the appearance of the image is that it looks sharper. A nice property of the function $f$ is that it produces values only in the

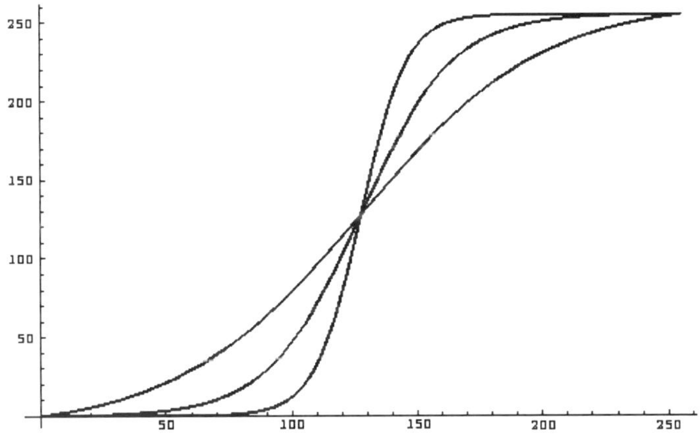


Figure 5.1: The sigmoid function $f(x)$ $\beta = 2, 1, 0.5$

range $[0 \ldots 255]$, irrespective of the input range. For example, if we convolve an image with a non-normalized kernel, the output range of this operation may no longer lie in $[0 \ldots 255]$; after the application of the sigmoid it does. The property that $f$ sharpens the image is especially useful

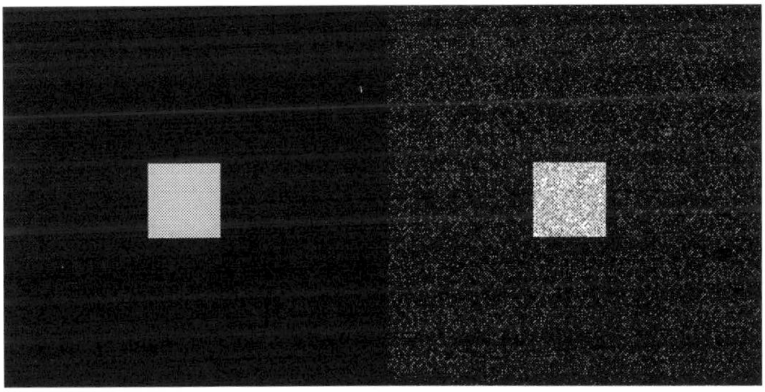

Figure 5.2: Test image without and with noise

for an application in low–pass pyramids. A low–pass filter blurs the image; the application of the sigmoid can reverse this blurring to a certain extent. Let us illustrate this on an example.

We have generated a black test image of size $256 \times 256$ containing a white square of size $50 \times 50$ (see Figure 5.2 left), and then we have added uniformly distributed random noise in the range $[-100 \ldots 100]$ to this image (see Figure 5.2 right). These test images are very simple (i.e., only edges parallel to the x- and y-axes) and are used to illustrate the basic concepts, in the second experiment we have used more realistic images.

We have built 4 levels of a $5 \times 5/4$ Gaussian pyramid on these two images (Figure 5.3 upper half; for display purposes, the images of the levels of the pyramids are zoomed to be of identical size, and no interpolation is used). In the case without noise, one can see that the edges of the square get blurred; this is due to the Gaussian filter. In the case with noise, the noise is gradually decreased from level to level. Then we have constructed the $5 \times 5/4$ sigmoid pyramid with $\beta = 2$ (Figure 5.3 lower half). One can see that the edges of the square are better preserved than in the Gaussian case. Also the noise is already reduced at the first levels of the pyramid. The black pixel on the upper right corner of pyramid level 4 is explained by the fact that only one pixel of the square is in the reduction window. This is a nice example of the shift variance of regular pyramid structures (see section 3.1) which is enhanced by the strong nonlinearity.

In a second experiment, we have used a $3 \times 3/4$ Gaussian pyramid on a real image. Figure 5.4 shows the Gaussian and the sigmoid pyramid ($\beta = 0.4$). We can see the same effect as with the white square. The Gaussian pyramid produces a blurred version of the image, and the borders of the objects get lost in higher levels. In the case of the sigmoid pyramid, the contrast is preserved, and objects can be detected more easily in higher levels. This effect is useful for image segmentation and threshold selection. For example, a threshold for the road can be easily found in the case of the sigmoid pyramid, but it is very hard for the Gaussian pyramid. This can also be seen from the histograms (Figure 5.5).

Of course, there are images where the application of the sigmoid pyramid introduces unwanted effects. As an example, consider an image with an unimodal histogram with the peak located in the middle of the histogram. In this case, the application of the sigmoid pyramid will result in an image with a bimodal histogram similar to Figure 5.5. The best results will be

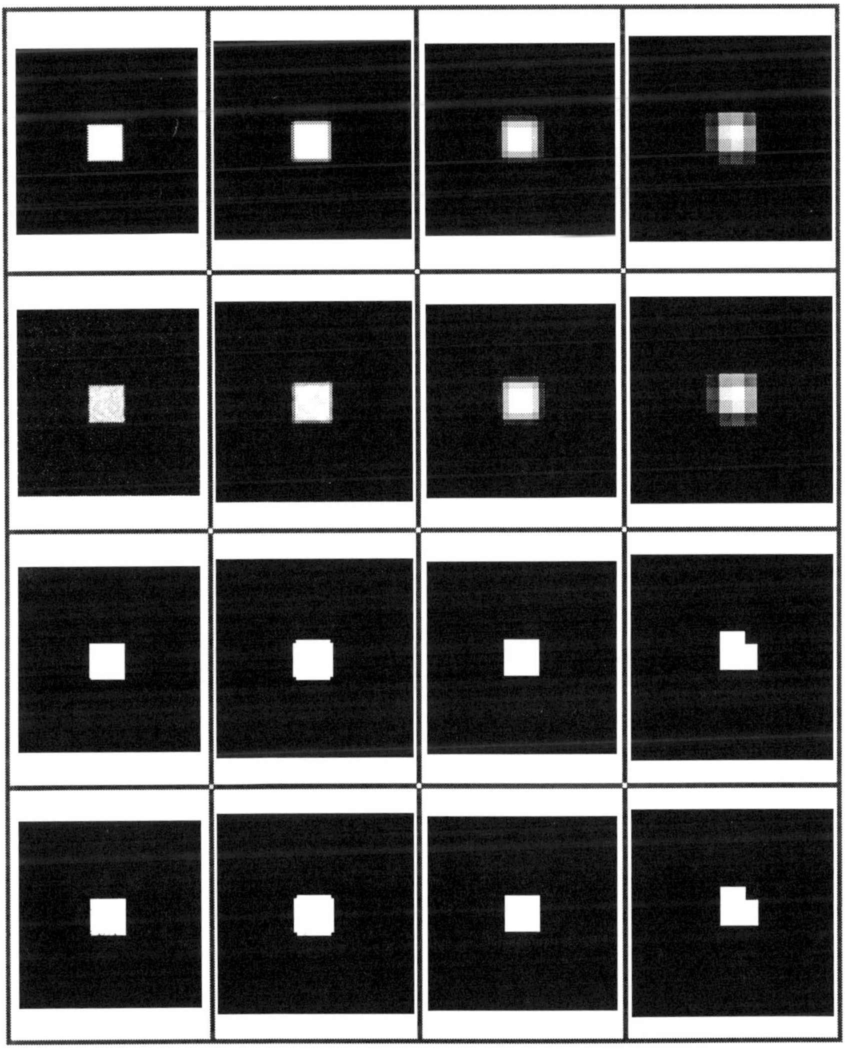

Figure 5.3: Gaussian (upper half) and sigmoid (lower half) pyramid of the test images

Figure 5.4: Gaussian and sigmoid pyramid of the road image

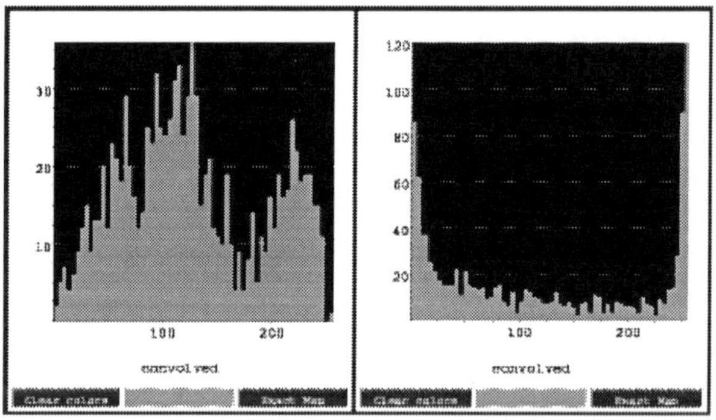


Figure 5.5: Histograms of level 3 of the Gaussian (left) and sigmoid pyramid (right)

obtained for images with bimodal histograms (i.e., dark or bright objects in the image).

A critical parameter of the sigmoid pyramid is $\beta$; $\beta$ controls the steepness of the sigmoid function as well as the "degree" of nonlinearity. If $\beta$ is chosen too large, we have saturation effects. The choice of $\beta$ depends on the image content and the goal of processing. For binary images, $\beta$ can be chosen large. This is also the case when the goal is to segment, an image because the nonlinearity enhances the differences between different regions. We could control $\beta$ by some adaptive procedure similar to adaptive threshold selection (Sahoo, Soltani, & Wang, 1988). Because a low–pass filter blurs edges, we can control $\beta$ by some edge operator (before low pass filtering). Whenever an edge detector yields an edge, we use a high value of $\beta$, whereas in regions with low or no gradient, we use a low value of $\beta$. The overall effect of this approach is similar to edge-preserving filters (e.g., Frost, Stiles, Shanmugan, & Holtzman, 1982; Lee, 1981; Nagao & Matsuyama, 1979).

### 5.1.2 Nonlinear Filters

We have already seen in the previous section how neural networks can be augmented by nonlinear processing from neural networks. But also other nonlinear filters (e.g., minimum or maximum) have been used as reduction functions for pyramids (Blanford & Tanimoto, 1988). Local maxima (minima) detection can also be performed by neural networks (Pao, 1989) by a mechanism called *Winner Takes All (WTA)*. Most proposed schemes have two disadvantages: first, the maximum is not preserved (i.e., the winning unit is always one), and secondly, the time for convergence depends on the number of units in the network. Recently Tsotsos (1991) introduced an algorithm for maximum selection that does not have these disadvantages. The idea is that the amount of inhibition between two units depends on the difference of their activities. Therefore, the unit with the maximum activation is never inhibited and preserves its value. The time of convergence depends only on the difference between the most and second most activated unit. With an additional parameter, the amount of iterations can also be controlled. Tsotsos used this mechanism in a hierarchical neural network to control the focus of attention. We can use it in the same way to build a maximum pyramid (for a minimum pyramid, we just have to reverse the signs).

### 5.1.3 Morphological Filters

Haralick et al. (1987) introduced the concept of a morphological pyramid. It has also been shown that certain types of neural networks (iconic neural networks) can implement morphological operations (Wilson, 1989). Indeed Wilson showed that these networks can compute also rank order filters (where maximum and minimum filters are special cases), which in turn can also be inferred from mathematical morphology (Serra, 1982). Wilson also showed that this class of networks can simulate grey-level morphology.

Other types of networks called morphological neural networks (Ritter, Li, & Wilson, 1989) use functions from mathematical morphology directly for updating the units. For our purpose, it is important to note that certain types of neural networks can be used to implement morphological functions, which in turn can be used to build morphological pyramids.

## 5.2 Symbolic Processing

We have seen in section 4.1 that we have to give up the one-to-one correspondence between cells and units in order to represent symbolic information; we have to replace a cell of a pyramid by a small neural network. We extend this to describe also processing of symbolic information by neural networks. We exemplify our ideas on the $2 \times 2/2$ curve pyramid, which has already been described in section 2.2.3.3. First, we build a neural network that performs the same operations as the $2 \times 2/2$ curve pyramid. Then we analyze how to generalize this network for a broader class of problems. Finally, we present some general ideas on neural networks and symbolic pyramids.

### 5.2.1 Neural Curve Pyramid

The $2 \times 2/2$ curve pyramid is an ideal test case to study neural network implementations of symbolic pyramids and to illustrate the concept of replacing a cell of a pyramid by a neural network. In order to simplify the discussion we do not describe end codes $F$ and U-turns, but they can be easily included.

A cell of the pyramid has four sides: North, East, West, and South; we denote them by $\Sigma = \{N, E, W, S\}$. Because a curve intersects a cell exactly twice, we have to represent binary relations among the sides; we call them *curve relations*. The curve is not oriented, therefore the relation is symmetric.

The representation of relations is as follows: Each cell has four output units $(N, E, S, W)$, which have connections to the cells in the next higher level. A relation is represented by activating exactly two units. For example, if unit $N$ and $S$ are activated, a North-South relation is represented. It is clear that, with this scheme, only one relation at a time can be represented unambiguously. Inside a cell, we represent the relations by a local representation as described in section 4.2 (i.e., one unit per possible relation). In order to perform the same reduction as in the $2 \times 2/2$ curve pyramid, we describe the following three steps in terms of neural networks (see section 2.2.3.3):

1. Split
2. Transitive Closure
3. Merge

#### 5.2.1.1 Split

The first operation to perform is to split the curve relations by the diagonal, as shown in Figure 5.6. A neural network to perform this operation is shown in Figure 5.7.

The neural network uses multiplicative connections (small circles with stars inside), for example, unit $ab$ gets activated only when $S$ and $E$ are activated. After multiplying the activations a unit sums the incoming signals and uses a threshold function $\Theta$ (with a threshold of 1) to compute the

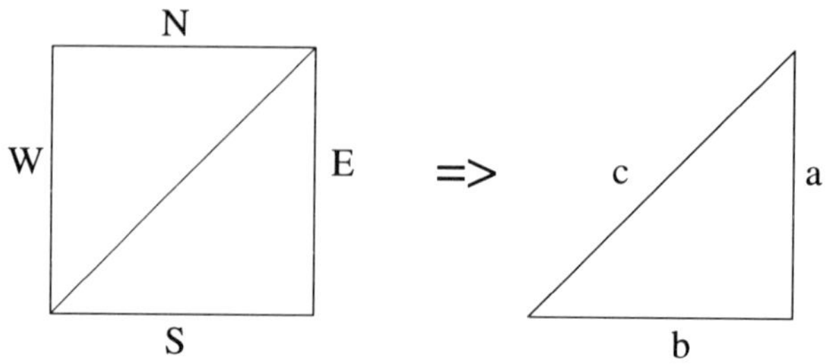


Figure 5.6: Splitting a curve relation by the diagonal

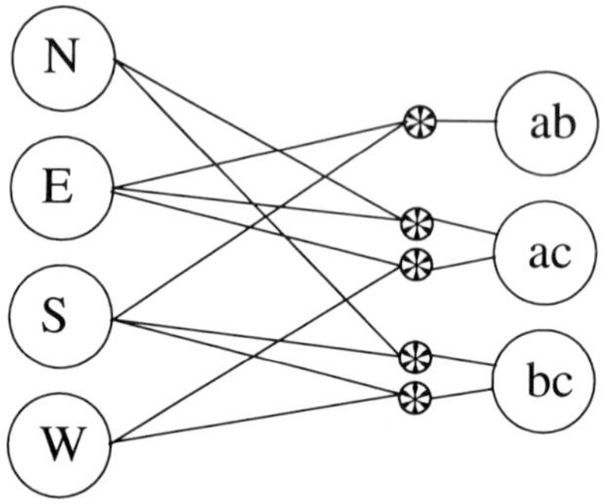


Figure 5.7: Neural network implementation of the splitting operation

output. For example, the activation of unit $ac$ is given by:

$$ac = \Theta(N * E + W * E)$$

We can also write this in form of a logical expression:

$$ac = (N \wedge E) \vee (W \wedge E)$$

or in terms of minimum and maximums operations:

$$ac = \max(\min(N, E), \min(W, E))$$

If only one relation is present and the activations of the units is 0 or 1, these formulations are equivalent. However, we see later the advantages and the interpretations of the various forms of operations.

One should note that three units are sufficient to represent all possible relations, because the relations are symmetric. The operation depicted in Figure 5.7 is performed for all four son cells.

#### 5.2.1.2 Transitive Closure

In the next step, we have to connect the curve segments in the four triangles by computing the transitive closure of the curve relations. We first merge the upper and the lower triangles in parallel (Figure 5.8a) and then merge the resulting two triangles to the final square, as depicted in Figure 5.8b.

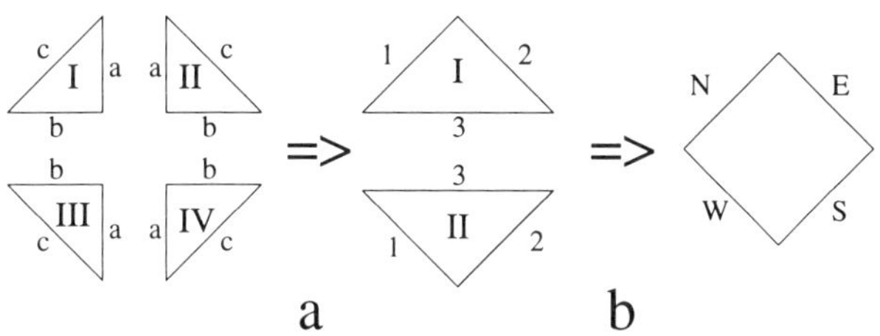


Figure 5.8: Transitive closure operation

The network implementation for the upper two triangles is shown in Figure 5.9(left) and for merging two triangles to a square in Figure 5.9(right).

Again small circles with stars indicate multiplicative connections and small arrows indicate direct connections. For example, we can express the activation of unit 13 as:

$$13 = \Theta(ac_1 * ab_2 + bc_1)$$

or equivalently:

$$13 = ac_1 \wedge ab_2 \vee bc_1$$

or as:

$$13 = \max(\min(ac_1, ab_2), bc_1)$$

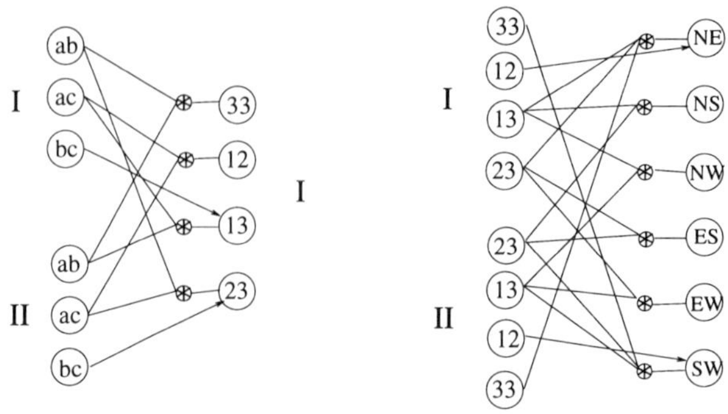


Figure 5.9: Neural network implementation of the transitive closure operation

#### 5.2.1.3 Merge

The third operation simply converts the local relation code used for processing the transitive closure to the code used by the output units. This can be done by the network shown in Figure 5.10. Here the units perform a simple summation (OR operation or MAX detection).

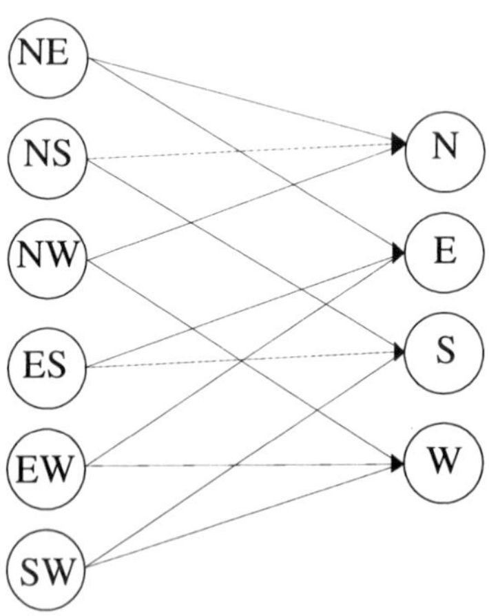


Figure 5.10: Neural network for performing the merge operation

When putting all these subnetworks together, we get a network for performing a reduction step for a single cell (see Figure 5.11). One can see that this network is again hierarchically structured and consists of many identical subnetworks. In order to handle U-turns and end-codes correctly, some additional units are necessary. In Figure 5.12, one can see four levels of a curve pyramid. Let us now analyze the properties of this network and discuss some possible extensions.

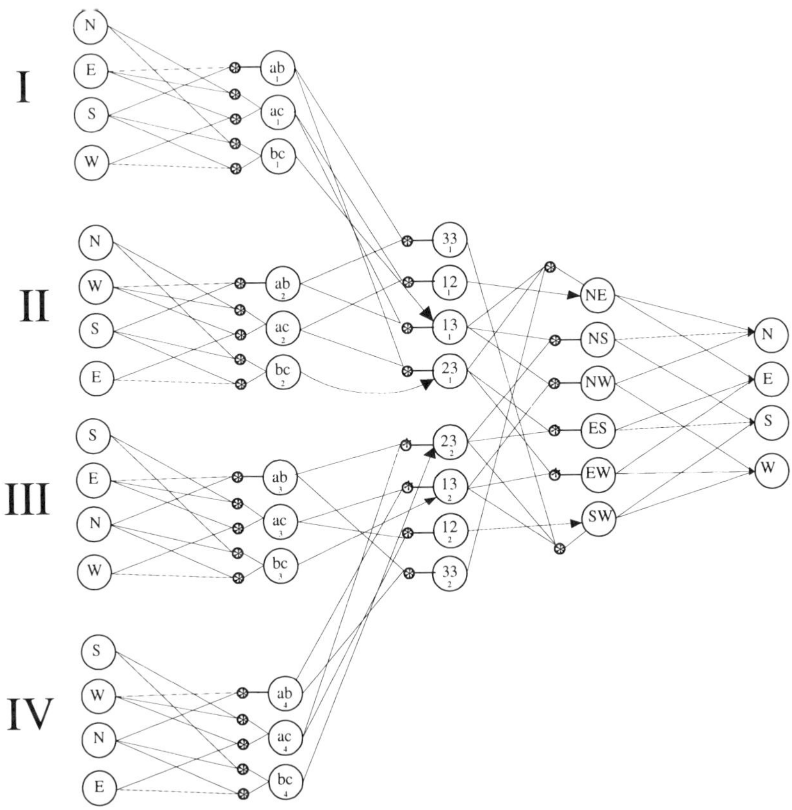


Figure 5.11: One cell of the neural curve pyramid

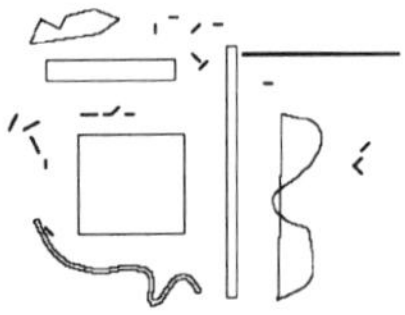

(a) Level 0

(b) Level 1

(c) Level 2

(d) Level 3

(e) Level 4

(f) Level 5

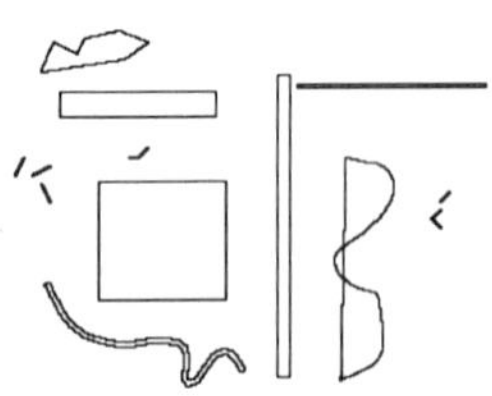

(g) Curve relations down projected from level 6 to original resolution

Figure 5.12: 6 Levels of a $2 \times 2/2$ curve pyramid; the images are rotated clockwise by 45 degrees

### 5.2.2 Properties of the Neural Curve Pyramid

From the construction of the neural curve pyramid, it is clear that it has the same properties as the $2 \times 2/2$ curve pyramid. That is, it possesses the length reduction property (i.e., the number of curve code elements decrease after every reduction step; see Kropatsch 1986, 1987a). This important property relates the highest level up to which a curve is still represented to the area that is traversed by the curve. This implies that short curves will disappear after a few levels, and only long curves will be represented in higher levels of the pyramid.

This property has been used for structural noise filtering (Kropatsch, 1986), which first builds the curve pyramid up to a certain level and then deletes all curves that are not represented in the levels below (see Figure 5.12g). The same kind of processing can be done in the neural curve pyramid. We have to add a circuit that is similar to the bottom up reduction.

With the neural curve pyramid, we are able to process curves with different "strength." For example, the curve relations can be generated by some edge detector that delivers a magnitude of the edge. In the previous sections, we have seen that we can write the individual steps in the bottom–up reduction as multiplications/additions, as AND/OR operations, or as MIN/MAX operations. AND/OR operations can only be used in the binary case, however, the other two are useful if we have a strength associated with a curve relation. Assume that the strength of a curve relation lies in the interval $[0 \ldots 1]$, where 0 means no relation and 1 is maximum. When we use the formulation with multiplications/additions, the curves at the higher levels will be represented only with a small strength. The formulation as MIN/MAX operations is more advantageous. We can get only values we have already at the base level. Also the interpretation of these operations is easy, because MIN and MAX are the fuzzy set operations for intersection and union for crisp sets (Betzdek, 1981). We can interpret the formulation with MIN/MAX operations as the fuzzy transitive closure of curve relations. Let us now investigate the fuzzy curve pyramid more closely.

### 5.2.3 Fuzzy Curve Pyramid

The binary curve pyramid as described previously has the disadvantage that it can only represent the presence/absence of a curve segment. This binary decision has several disadvantages:

- We must decide already at the base level about the presence/absence of a curve segment. This typically involves the determination of a threshold for an edge detector. This is a problematic step, because either we miss curve segments that cause disconnected curves (therefore the curve will disappear after a few levels) or we add too many segments that might connect otherwise disconnected curves.
- Another problem with the binary representation of curves arises when different curves meet in one pyramidal cell. This might cause ambiguities in the representation due to the restricted storage capacity of the pyramidal cells.

Therefore, it would be advantageous to have a non-binary representation of curves in a pyramid. Because a curve introduces a relation between two sides of a pyramidal cell, a straightforward generalization is to consider fuzzy relations instead of binary ones. The strength of the curve is represented by the grade-membership of the fuzzy relation.

**Definition 6 (Fuzzy Relation)** *Let $X = \{x_1, \ldots, x_k\}$ be a $k-$element fuzzy (or crisp) set. The set $R \subseteq X \times X \times [0 \ldots 1]$ is a fuzzy relation. We call the triple $(x_i, x_j, \mu(x_i, x_j))$ $(x_i, x_j \in X, 0 \leq \mu(x_i, x_j) \leq 1)$ a fuzzy relation between $x_i, x_j$ with grade-membership $\mu(x_i, x_j)$.*

We say for $\mu(x_i, x_j) = 0$ that there is no relation between $x_i$ and $x_j$. The relation is symmetric if $\mu(x_i, x_j) = \mu(x_j, x_i)\ \forall x_i, x_j \in X$. In what follows, we always assume symmetric relations.

The transitivity of a fuzzy relation is expressed as:

$$
\begin{aligned}
&(aRb, \mu(a,b)), (bRc, \mu(b,c)) \Rightarrow \\
&\qquad (aRc, \mu(a,c) = min(\mu(a,b), \mu(b,c)))
\end{aligned}
\tag{5.5}
$$

#### 5.2.3.1 Properties of the Fuzzy Curve Pyramid

We have described the individual steps necessary to construct level $n+1$ from level $n$ of the fuzzy curve pyramid. In this section, we show several properties of the fuzzy curve pyramid.

**Property 1 (Length Reduction)** *The fuzzy curve pyramid has the length reduction property (i.e., it reduces the number of curve code elements after every reduction step) similar to the $2 \times 2/2$ binary curve pyramid.*

**Proof** The proof is essentially the same as for the binary case (i.e., all curve relations with grade-membership $\mu > 0$ have to be considered). One can convert fuzzy curve relations to binary ones by setting an arbitrary threshold $\Theta$ at the grade-membership $\mu$.

**Property 2 (Minimum Strength)** *If a single curve intersects the receptive field of a cell at level $n$, the grade-membership of the relation at level $n$ will be the minimum of the grade-memberships of the curve segments of the base level.*

**Proof** Because we are dealing with a single curve, we can prove this property recursively; therefore, we need to consider only two successive levels of the pyramid. We can consider each individual step in the pyramid construction separately. In the case of a single curve, the split and merge operations are simple copying mechanisms; therefore no changes in the grade-memberships occur. For this reason, we have only to check the transitive closure operation. From the definition of the fuzzy transitive closure, one can see that if two segments are connected, the grade–membership of the resulting segment is assigned the minimum of the grade–membership of the individual segments. If we apply this recursively, property 2 follows.

Note that property 2 does not necessarily hold for the case when two or more curves pass the receptive field of a cell. However, in this case, we can prove a more general property: If we define the strength of a curve as the minimum strength of its segments (see above), then the strength of a fuzzy curve relation $AB$ is the maximum strength of all curves connecting the side $A$ with side $B$. This allows us to formulate the following property:

**Property 3 (Maximum Strength)** *Let $C$ be a curve in the base of the fuzzy curve pyramid with strength $\mu(C) = \omega_0$. Then this curve will remain connected by the fuzzy reduction process and the strength of all fuzzy curve relations derived from $C$ will be greater to or equal to $\omega_0$, that is, the strength of the curve is preserved.*

**Proof** 1. Assume that $C$ has the greatest strength of all curves in the image. In this case, taking the maximum at all operations where curve segments are merged preserves the strength as long as the curve does not disappear.

2. If the strength of $C$ is not maximum, it could happen that it gets merged with a curve of higher strength. The combined segment will receive the higher strength, for example, $\mu(s_1 \cup s_2) = \max(\mu(s_1), \mu(s_2)) \geq \mu(s_1)$. qed.

The two last properties can be combined in the *minmax principle*, which is satisfied by the fuzzy curve pyramid.

**Property 4 (Minmax Principle)** *The strength of a curve is the minimum of the strength of its segments. And the strength of a side of a cell in the curve pyramid has the maximum strength of all curves intersecting this side.*

#### 5.2.3.2 Enhancements

Until now we have not fully exploited the information provided by the grade-membership of the relations. This is discussed in this section.

**5.2.3.2.1 Ambiguity Reduction** The basic idea is that due to the grade-membership of the relations, we can discriminate between curves of different strength. This is important when more than one curve meets in a single cell. Consider as an example Figure 5.13. We see that two curves with different strengths meet in a common cell. In the case of the binary

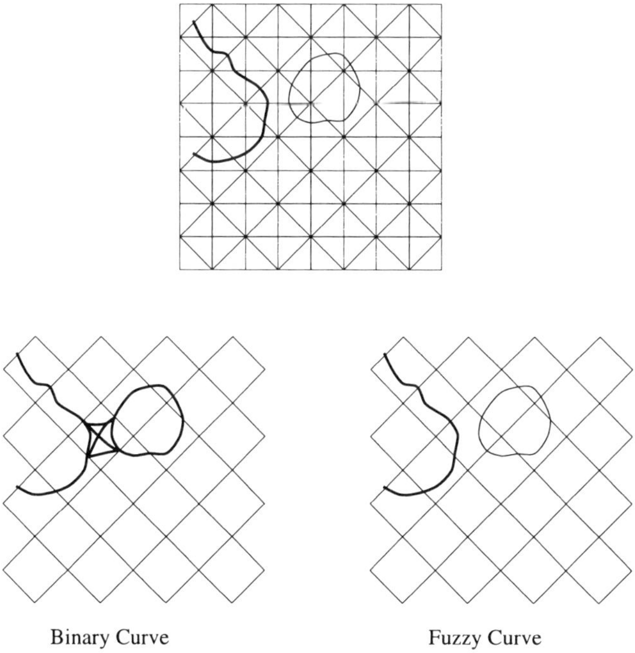


Figure 5.13: Example where curves get merged in the binary curve pyramid and remain separate in the fuzzy curve pyramid

curve pyramid, we cannot discriminate them; therefore, we introduce all possible relations, and the two curves are merged. Whereas in the case of the fuzzy curve pyramid, we know due to the grade-membership that there are two curves (a strong one from North to West and a weak one from South to East).

There are two possible ways to modify the algorithm to introduce this ambiguity reduction:

1. **Modification of the Split Step:** We can modify the split step in order to take the different grade-memberships into account. However, this does not eliminate all ambiguities; for example, curves with the same strength, two curves entering at the same side of a cell.

2. **Eliminate Merge and Split:** Before the merge step, we have all the necessary information about the curve relations. If we eliminate the merge and split step, we do not introduce ambiguities. This can be done in the following manner: A single cell in a $2 \times 2/2$ pyramid has two fathers (see Figure 5.14A). Therefore, we can connect the result of the transitive closure operation directly to the two fathers of a cell, in the way it is depicted in Figure 5.14B, that is, we eliminate the merge step and perform the split operation. Figure 5.14C,D show an example of how the curve relations are transferred to the fathers. One can see that only those nodes get activated where a curve is present. The disadvantage of this operation is that we need 10 links between the cells, whereas we need only 8 links for the original algorithm; however, the number of links (processing steps) in a cell is reduced. It is important to note that as long as the curves are not parallel (entering and leaving at the same sides), they can be discriminated. In the case of parallel curves, they will get merged. The resulting activation of the new curve segment is the maximum of the activations of the two parallel segments.

It should be noted that the methods we have proposed in this section do not alter the properties we have discussed previously.

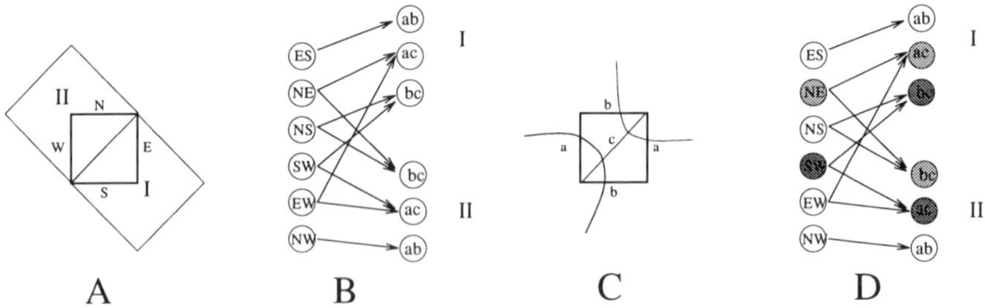


Figure 5.14: Direct connection between the cells (without merge step)

**5.2.3.2.2 Curve Selection** The binary curve pyramid extracts two important properties of curves, that is, length and connectedness. The same properties are extracted with the fuzzy curve pyramid but has additionally the curve strength as property (see property 2). We can use this additional information to select certain curves in the pyramid construction or down projection phase. For example, we can build the pyramid up to certain level, then remove all curve relations with a grade-membership below a threshold. If we then down project the remaining relations, we get a structural filter that also takes the curve strength into account.

#### 5.2.3.3 Evaluation

The method we have presented shares many similarities to Wu and Leahy (1993). As one of our next steps, we would like to consider the inclusion of the maximum flow algorithm (Gomory & Hu, 1961) to process fuzzy relations. The minimax principle of curves follows a similar objective as active contour models (e.g., Snakes; Kass, Witkin, Terzupoulos, 1987). It combines the cross sectional strengths along the path of the curve. However, unlike snakes, the minmax principle uses min and max operators to combine "internal" and "external" forces instead of a linear (weight) combination. A lot of research has been done using hierarchical representation of edge information (i.e., edge pyramids; Park, Park, Lee, & Choi, 1994; Schrift, Zeevi, & Porat, 1988). However, these approaches store numerical information in the pyramidal cells, whereas the fuzzy curve pyramid combines symbolic with numeric information.

Compared to the binary curve pyramid, the fuzzy curve pyramid needs more storage space. For the $2 \times 2/2$ curve pyramid, we need only 8 bit per cell to represent the curve relations. For the fuzzy curve pyramid, we need $8 \times 8$ bit per cell (if the end relations are also considered). The advantage gained is that we can process curves with different strengths. Besides the storage requirements, the computational complexity of the binary and the fuzzy curve pyramid are the same, that is, we can build both pyramids in a logarithmic number of steps on a parallel hardware.

### 5.2.4 Distributed Curve Pyramid

In section 4.3, we have seen how we can learn a distributed code for relations by an auto-associative network. However, we need also to process these relations. This cannot be done the same way we did it for local representations of the relations, where we have designed a network by hand. Therefore, we have to employ some learning algorithm. A simple way would be to enumerate all possible combinations of relations and training of a suitable network by a supervised learning method (e.g., back-propagation). The result would implement the desired reduction function with a distributed representation. With this approach, no new representations (e.g., new curve primitives) will arise.

For these reasons, we would like to employ some unsupervised learning method that has the freedom to also form new representations. A possible architecture for this goal is similar to the auto-associative network we have used for learning the distributed curve relations.

However, with the unsupervised method, we are facing the fundamental problem that the code developed at the output is not necessarily the same as that used as an input. Therefore, we have to train all levels of the pyramid and use different reduction functions at these levels. The other problem is that we cannot decode the distributed code into a local one for interpretation purposes. In order to achieve the desired properties, there is still a lot of research necessary. We discuss in chapter 6 some possible learning schemes that might be useful also for symbolic pyramids with distributed representations.

### 5.2.5 Neural Networks for Symbolic Pyramids

Though the scheme with the distributed representation should be more general than the network with local representation, it is not general enough for all kinds of symbolic pyramids. Replacing a cell of a pyramid by a small neural network gives us enough freedom to perform more general computations. In order to process symbolic information, we need a network that can act as a finite state machine.

In general, these small networks can be considered as grouping small image parts to objects (or parts of objects). Certain constraints have to be satisfied by this grouping process (e.g., the parts meet at a certain angle). Because the grouping is done hierarchically, the "combinatorial explosion" of checking all possible combinations can be avoided. Moreover, we can use learning algorithms to alter this grouping process. From these considerations, we can describe the required properties of the networks.

1. *Recurrent network* - each state of the network can be associated with a symbol.
2. *Constraint satisfaction* - the network must converge to a stable state that fulfills as many constraints as possible in order to arrive at a "good" interpretation of the scene.
3. *Distributed representation* - as we have seen in section 4.3, an efficient representation of symbolic structures has to be distributed.
4. *Unsupervised Learning* - if learning should be performed by the small networks, it has to be unsupervised in order to avoid the problem of teaching each cell/level of the pyramid separately (Fukushima et al., 1983; Yamaguchi & Kropatsch, 1989).

There are some network models that fulfill the properties 1 - 3, for example, Hopfield network (Hopfield, 1982, 1984), recurrent Jordan network (Jordan, 1986), and Boltzmann machine (Hinton & Sejnowski, 1986). But all these models use a supervised learning scheme. Some research has to be done to find an ideal combination for our purposes. Especially the learning problem is still a major obstacle to perform these tasks.

## 5.3 Summary

In this section, we have analyzed the processing of image pyramids. In the case of numeric processing, we have seen that we can build the same reduction functions with neural networks. We have also analyzed the role of sigmoid functions in pyramidal processing and have seen that the resulting *sigmoid pyramid* has some interesting properties in terms of preserving object contours and noise removal.

In the case of pyramids with symbolic content in the cells, we have analyzed the $2 \times 2/2$ curve pyramid and have presented the corresponding *neural curve pyramid*. We have seen that with this formulation, we can also process curve strength. Indeed we have shown that this type of pyramid has an interpretation in terms of fuzzy set theory, therefore, we have called it *fuzzy curve pyramid*. For the fuzzy curve pyramid, we have proved several important properties.

Then we have analyzed the processing of symbolic information in pyramids in more general terms and have seen that learning is still a major problem for neural networks dealing with this kind of information.

In the last three chapters (Structural Considerations, Contents and Representation, and Processing), we have shown that we can build for any type of pyramid an equivalent neural network. We have shown by several examples that through description of a pyramid in terms of neural networks, we can enhance pyramids. Until now we have not considered learning a fundamental function of most neural networks in detail. The next section shows how to incorporate learning algorithms in pyramids.

# Chapter 6

# Learning Pyramids

In the previous chapters, we have seen that hierarchical neural networks and image pyramids have a lot of things in common. This is especially true for grey-level pyramids. However, one major difference between these models still remains. For neural networks, there exists a variety of learning algorithms. This is not the case for image pyramids.[1] With the knowledge about the similarities between these models, we can employ learning algorithms developed for neural networks also to image pyramids. In particular, we would like to study learning algorithms for regular, grey-level pyramids that use convolution operations as reduction functions. The results of this chapter can, to some extent, be generalized to other types of pyramids.

## 6.1 Learning Algorithms for Pyramids

What is the reason for studying learning algorithms in pyramids? The Gaussian kernel has some unique properties (Yuille & Poggio, 1986) and is therefore widely used as a reduction function in pyramids. The assumption underlying these results is that all possible images have to be processed. But for a specific application, one has to deal with a much smaller class

[1]Pyramid (re)linking might be considered as a form of learning.

of images. For this reason, it might be advantageous to have an adaptive procedure that can tune the kernel for a specific application. A learning algorithm for pyramids should possess the following properties:

1. An ideal learning algorithm for pyramids should be *unsupervised*, in order to avoid the problem of telling each cell which value it should adopt for a given input. This avoids also the problem of long learning times and bad scaling characteristics of most supervised learning algorithms (Hinton, 1989).

2. We also want to keep the main properties of pyramids, that is, a *parallel* and *local* implementation of the learning algorithm must be possible.

3. We further want each cell, or group of cells, to perform the *same reduction function*.

4. Of course, learning should be *fast* and require only a few iterations.

5. The learning algorithm should be incremental so that the pyramid can *adapt to gradual changes* or drifts in its environment.

Because a pyramid reduces the information content of an image level by level, an objective for the reduction function would be to *preserve as much information as possible*, given the restrictions imposed by the structure of the pyramid. Or stated the other way round, to minimize the information loss by the reduction function. This criterion is independent of a specific application, however, very useful for pyramidal algorithms. Usually a pyramid is used as a data structure for images where algorithms operate on. A large class of algorithms proceed top–down. If enough information about the image is already available at the low resolution, the high resolution parts of the image need not be processed. Therefore, a learning algorithm that preserves as much information as possible would be beneficial for these cases and reduce the overall processing time.

A technique known from statistics to reduce the dimensionality of data while preserving valuable information is *Principal Component Analysis*

*(PCA)*, also known as Karhunen-Loéve expansion (Anderson, 1958). For this reason, we would like to examine this technique in more detail and see if we can employ it in a pyramidal framework. To be more precise:

Let the input to a cell in a $n \times n/r$ pyramid be denoted by the vector $\xi \in \mathbb{R}^{n^2}$ (this vector is formed by all grey-values of the sons) and let $w \in \mathbb{R}^{n^2}$ be a vector of the weights of the kernel. A pyramidal cell calculates:

$$\eta = \sum_{j=1}^{n^2} w_j \xi_j = w^T \xi = \xi^T w \tag{6.1}$$

where $\xi_j$ and $w_j$ is the j-th component of $\xi$, and $w$ and $\eta$ is the output of the cell.

Assume that the vectors $\xi$ are drawn from some probability distribution $P(\xi)$. We can calculate the correlation matrix $C$ by:

$$C_{ij} = \langle \xi_i \xi_j \rangle = \langle \xi \xi^T \rangle \tag{6.2}$$

where $\langle \cdot \rangle$ indicates the average over the input distribution. If $\xi$ has zero mean $\mu_i = \langle \xi_i \rangle = 0$ for all $i$, then $C$ is the covariance matrix of the input distribution. We want $w$ to lie in the maximal eigenvector direction of $C$, that is, $Cw = \lambda w$, where $\lambda$ is the largest eigenvalue of $C$. In this case, $w$ is the first principal component of $P(\xi)$ and maximizes the variance of $\eta$, that is, $\langle \eta^2 \rangle \rightarrow max$. Moreover, we want $w$ to be normalized, that is, $|w| = 1$.

We have to look for an unsupervised, local learning algorithm that is capable of extracting the principal components of the input data. Indeed several such algorithms have been proposed. We briefly review them and evaluate them according to our needs.[2]

## 6.2 Principal Component Analysis Network

The first person who proposed a unit that is capable of extracting the first principal component of the input distribution was Oja (1982). He used a

[2]A recent review on learning algorithms in linear networks can be found in Baldi and Hornik (in press).

linear unit, that is,

$$\eta = \sum_i w_i \xi_i \tag{6.3}$$

and the following learning rule:

$$w_i(t+1) = w_i(t) + \gamma(t)\eta(t)[\xi_i(t) - \eta(t)w_i(t)] \tag{6.4}$$

where $\gamma(t) \ll 1$ is a learning factor that should be slowly decreasing to zero (e.g., $\gamma(t) \sim \frac{1}{t}$). Writing equation 6.4 in another form

$$w_i(t+1) = w_i(t) + \gamma(t)\eta(t)\xi_i(t) - \gamma(t)\eta^2(t)w_i(t) \tag{6.5}$$

one can see that the second term of the right hand side is the usual Hebb rule (Hebb, 1949), and the last term is a control term that prevents the weights from growing. It can be proved that the weight vector $w$ has unit length, that is, $|w| = 1$ or $\sum_i w_i^2 = 1$, this is due to the third term in equation 6.5. Oja also proved that $w$ lies in the direction of the maximal eigenvector of the correlation matrix (covariance matrix for zero mean data). It is surprising that such a relatively simple learning rule can achieve such a complex task.

There are other related methods for principal components extraction, for example, Yuille Kammen, and Cohen (1989) suggested following rule:

$$w_i(t+1) = w_i(t) + \gamma(t)[\eta(t)\xi_i(t) - w_i(t)|w(t)|^2] \tag{6.6}$$

The direction for $w$ is the same as in the case of Oja's rule, the length of $w$ is $|w| = \sqrt{\lambda_{max}}$ (where $\lambda_{max}$ is the maximal eigenvalue of $C$). The advantage of this rule is that it has an associated cost function that is minimized in a gradient descent way by equation 6.6, however, the learning rule is not strictly local.

In a series of papers, Linsker (1986) studied another type of rule, where he used explicit clipping of the weights (i.e., $w_- \leq w_i \leq w_+$). He showed that a layered network of units trained with this learning rule produces center-surround cells and in higher layers, orientation selective cells, although the input consist only of independently distributed random noise. Linsker proposed a general principle (Linsker, 1988) called the *infomax principle*, which states that a layer of a network should be trained to preserve the maximum of information of the previous layer. He also showed in this paper that in the case of a Gaussian distribution of the input signal, principal component extraction is equivalent to maximizing the information content of the output signal.

Until now we have talked only about learning rules that are able to extract the first principal component. Other rules have been proposed that can extract the first $M$ principal components. Among them Sanger (1989) suggested a two–layer network with $M$ output units in the second layer for extracting the first $M$ principal components. The output of unit $i$ is given by:

$$\eta_i = \sum_j w_{ji}\xi_j = w_i^T \xi \tag{6.7}$$

and the learning rule of Sanger is:

$$w_{ij}(t+1) = w_{ij}(t) + \gamma(t)\eta_j(t)[\xi_i(t) - \sum_{k=1}^{i} \eta_k(t) w_{jk}(t)] \tag{6.8}$$

Oja (1989) proposed a similar rule:

$$w_{ij}(t+1) = w_{ij}(t) + \gamma(t)\eta_j(t)[\xi_i(t) - \sum_{k=1}^{M} \eta_k(t) w_{jk}(t)] \tag{6.9}$$

The only difference is in the upper limit of the summation. Sanger's rule finds the exact first $M$ principal components, whereas in the case of Oja's $M$-unit rule, only the space of the first $M$ principal components is spanned, and the output units are not decorrelated. In fact, the output units will have almost equal variance. Sanger and Oja applied their rules to image compression and texture segmentation.

Another type of network that is able to extract principal components is a three–layer feed–forward network trained to auto-associate the input with the output. We have already encountered this type of network in section 4.3, where we have used it to learn a distributed code for relations. This network consists of $N$ input and output units and $M < N$ hidden units and is trained (usually by back-propagation) to reproduce the input at the output units. Bourlard and Kamp (1988) analyzed this type of network and found that nonlinearities in the hidden units are not necessary.

They proved that the hidden units project the input on the space spanned by the first $M$ principal components of the input distribution. Another interesting result in this paper is that the inclusion of a bias term (see equation 2.2) in the hidden and output units reduces the problem to zero mean data (therefore no explicit normalization is required). The network produces essentially the same result as Oja's $M$-unit rule, with equal variance on average on each of the hidden units. These results have been extended by Baldi and Hornik (1989). They showed that the usual quadratic error function $\mathcal{E}$ (see equation 2.8) has a unique global minimum corresponding to the projection onto the subspace generated by the first $M$ principal component vectors. All other critical points of $\mathcal{E}$ are saddle points. For this reason, $\mathcal{E}$ has no local minima, and back-propagation can be expected to converge (although it might get stuck in a saddle point).

Other types of multilayer networks with lateral inhibition in the hidden layers have been proposed (e.g., Kung & Diamantaras, 1991). These networks use the lateral connections to decorrelate the hidden units, so that the hidden units will extract the "real" principal components.

In a very recent paper, Xu (1993) proposed another type of network for principal component analysis. His network is a layered network with bidirectional symmetric links between the layers. This work is interesting for several reasons. First, it embeds most of the known other networks for principal component extraction. In particular, the Oja learning rule is a special case of this network, and the learning equations are very similar to the auto-associative back-propagation network. Secondly, Xu showed that this network converges to the exact principal components when different amplification terms are included in the output units. Xu also studied the effect of using sigmoid functions in the units. They act as a symmetry breaker, so that the network converges to the exact principal components.

In a recent paper, Hornik and Kuan (1992) analyzed the convergence properties of these algorithms,[3] by studying the differential equations that govern the learning algorithms. The results of this analysis are that from the point of stability, the multilayer versions of the principal component

[3]The algorithm of Xu was not included in this study.

networks are preferable. Also the speed of convergence should be better.

Let us now examine how we can incorporate these networks and learning rules in a pyramidal framework.

## 6.3 Principal Component Pyramids

### 6.3.1 Oja Rule

Let us first examine the Oja one–unit rule. It is fairly obvious how to implement it in a pyramid. We start with some kernel (e.g., a Gaussian, random, etc.). We present an input image at the bottom level of the pyramid and compute the reduced resolution. Then we apply one update step of the learning rule (6.4) to all cells and average the weight change of all cells. We repeat the process until the kernel has converged. Then we apply the same procedure to the next level of a pyramid with a new kernel, and so on.

A few comments are necessary: First, as we have described the learning rule here with the averaging over all weight updates, the learning rule is no longer local, because each cell has to communicate its weight change to all other cells on the same level. However, we do this only for convenience, because in a real pyramid architecture, a cell could maintain its own set of weights, and if we present many images that have the same local statistics, all cells will end up with the same (or nearly the same) kernel.

Another comment concerns the usage of different kernels at different levels of the pyramid, because usually in a pyramid there is only one kernel for all levels. To use more than one kernel is important, because the local statistics may change from level to level and therefore different kernels are appropriate. Consider a simple example of an overlapping pyramid where the input at the base level is independently distributed random noise. The correlation matrix will be the identity matrix. And the weights of the kernel will be all positive and identical, so that a cell performs a simple averaging. But at the next level, we have due to the overlap of the pyramid a high correlation between nearby cells, and therefore the weights will develop a Mexican hat type kernel. So in order to maintain the property of information preservation, it is important that we use different kernels at different levels.

If we would like a pyramid that is adapted to a set of images, we could use the same procedure as previously described, but make only one update step with every image and repeat the presentation of the images.

### 6.3.2 Oja Rule with Maximum Detection

Images have often large homogeneous regions. Because these regions do not contain very much information, we can change the update procedure a little bit, by letting only those cells change their weights that are local maxima. We have already seen in section 5.1.2 that a simple network proposed by Tsotsos (1991) can perform local maxima detection and preserves the value of the maximum. All other cells that are not local maxima will become 0. Because the learning equation 6.4 does not change the weights if the output of a cell is zero, cells at nonlocal maxima positions will not learn anything. This will happen especially in homogeneous regions, so that the kernel will not adapt to these regions.

### 6.3.3 Auto-associative Networks

The Oja rule works for all kinds of regular pyramid structures, however, for overlapping pyramids, we do not eliminate all kinds of redundancies. Because two neighboring cells will have a part of their input in common, they will represent the overlapping part twice, and their activity will be correlated. Therefore, we do not preserve as much information as possible. This effect is simply due to the fact that every cell has the same reduction function and is trained independently of the others. In order to avoid this, we have to give up the restriction that every cell computes the same reduction function. Because we do not want each cell of a pyramid to compute a different function, we allow only a small number of different cell types. This number is determined by the structure of the pyramid, as we show later. With this extension, we can allow the cells to compute more than the first principal component. Therefore, more information will be preserved.

We have to take a learning rule that is capable of computing more (say $M$) principal components. Because the analysis of Hornik states (Hornik & Kuan, 1992) that the solution of the auto-associative networks are more stable than those of the two–layer networks, we use this type of learning rule. A further advantage is that the auto-associative network automatically generates also the expansion function of the pyramid level $n$ to level $n-1$. The fact that these networks do not produce the first $M$ principal components and span only the space of the first $M$ principal components is an advantage in our case, because we do not want the cells of a pyramid to become ordered. Because the hidden units have equal variance on the average, everyone of them is of equal importance, and this should in turn facilitate the robustness of the procedure. In order to apply such networks to a pyramid structure, we have to satisfy several constraints:

We have to build a three-layer network that has as input a part of level $n$ of the pyramid. The hidden layer represents a part of cells at level $n+1$, and the output layer represents cells of the expanded level $n'$. It should be noted that we do not use full connectivity between the layers; we use the same connectivity as in the pyramid. This is similar to the Laplace Pyramid (Burt & Adelson, 1983), where an image is reduced by a reduction function $R$ (corresponds to the transformation from the input units to the hidden units) and then expanded with an expansion function $E$ (corresponds to the transformation from the hidden units to the output units).

The goal is to train the network such that the reconstruction error for the whole level is minimized (i.e., $||L_n - L_{n'}||^2 = ||L_n - E(R(L_n))||^2 \rightarrow \min$). This minimization is achieved by back-propagation training.

The simplest case are non-overlapping pyramids, for example, the $2 \times 2/4$ pyramid. We have one cell in the hidden layer and four cells both in the input and the output layer, and the result will be the same as in the case of the Oja rule. But with a back-propagation network, we can also take $4 \times 4$ cells as input and output and four cells in the hidden layer. Because there is no overlap, the cells will develop identical weights, so there is no use in taking a larger input for non-overlapping pyramids.

The interesting cases occur with overlapping pyramid structures.

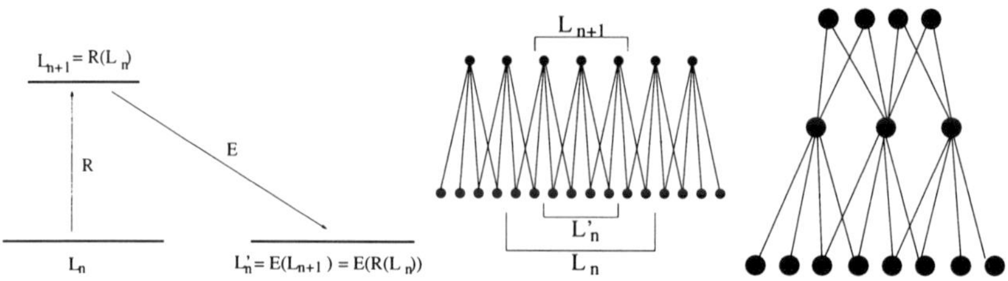


(a) General Setting (b) 4/2 pyramid (c) Resulting network

Figure 6.1: From the structure of the pyramid to the topology of the network

Figure 6.1 depicts the one-dimensional situation. Figure 6.1a shows the general goal to be achieved and the notations employed; Figure 6.1b shows a 4/2 pyramid. When constructing the corresponding network, we start at the output layer (i.e., $L'_n$). For an $n \times n/r$ pyramid, we typically choose the size of the output layer as $n \times n$. Next, we have to include all fathers of the cells in the output layer as hidden units. Finally, we have to include all sons of the hidden layer cells in the input layer. For the 4/2 pyramid, this results in an $8 - 3 - 4$ network as shown in Figure 6.1c.

In the two dimensional case let us first examine the case of the $4 \times 4/4$ pyramid. This pyramid has the nice property that every cell has the same number of fathers; this is not the case for all other structures.

Figure 6.2 shows two levels of a $4 \times 4/4$ pyramid with the circles marking the position of the cells in the higher level. Also shown are the receptive

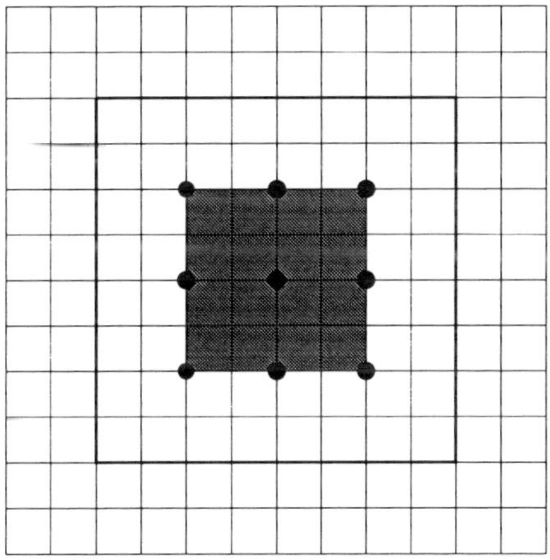

Figure 6.2: Two levels of a $4 \times 4/4$ pyramid

fields of the 9 higher level cells. In designing an auto-associative network for this pyramid, we start by specifying the number of output units. In this case, it is the square of $4 \times 4$ cells (shaded grey). Because these cells correspond to the expanded cells of level $n$, we have to include all cells of level $n+1$ in the hidden layer that are fathers of these cells, in our case, $3 \times 3$ cells. At the input layer, we have to include all sons of the $3 \times 3$ cells in the level $n$, in our case, $8 \times 8$ cells. So the topology of the network is: $8 \times 8 \mapsto 3 \times 3 \mapsto 4 \times 4$.

The next thing we have to look at are the constraints of the weights of this network. Because the expansion must cover the whole level $n$, the hidden units overlap, therefore, we have only 4 independent hidden units. Figure 6.3a depicts which hidden units are forced to develop identical weights (same letter). Because the expansions do not overlap, we have no constraints at the weights of the output units.

The same kind of analysis can be done for different pyramid structures in order to find the topology of the corresponding network. Table 6.1 lists several of the resulting topologies for commonly used pyramid structures.

The constraints we have listed are due to the overlap in the pyramids. Of course, implementing them in a pyramid structure the constraints are

(a) (b) (c)

Figure 6.3: Constraints for the weights of a $4 \times 4/4$ (a), $5 \times 5/4$ (b) and a $2 \times 2/2$ (c) pyramid; same letters indicate the same set of weights

| structure | network | | | constraints |
|---|---|---|---|---|
| $2 \times 2/4$ | $2 \times 2$ | $\mapsto 1$ | $\mapsto 2 \times 2$ | no constraints |
| $3 \times 3/9$ | $3 \times 3$ | $\mapsto 1$ | $\mapsto 3 \times 3$ | no constraints |
| $3 \times 3/4$ | $7 \times 7$ | $\mapsto 3 \times 3$ | $\mapsto 5 \times 5$ | Hidden and output units constraint as shown in Figure 6.3a |
| $4 \times 4/4$ | $8 \times 8$ | $\mapsto 3 \times 3$ | $\mapsto 4 \times 4$ | Hidden units constraint as shown in Figure 6.3a |
| $5 \times 5/4$ | $13 \times 13$ | $\mapsto 5 \times 5$ | $\mapsto 7 \times 7$ | Hidden units constraint as shown in Figure 6.3b |
| $2 \times 2/2$ | $4 \times 4$ | $\mapsto 5$ | $\mapsto 2 \times 2$ | Hidden units constraint as shown in Figure 6.3c |

Table 6.1: Different Network Topologies for Different Types of Pyramids

automatically satisfied by the arrangement of cells, therefore, learning is still local in the group of cells.

The thus constructed network can be trained by some suitable learning algorithm, typically of the back-propagation type, using batches of an image as input for training the first pyramid level. After that, the second level of the pyramid can be trained in the same way using the first pyramid level as training data, and so on. In order to avoid border effects, the input image is assumed to be symmetric with respect to the horizontal and vertical borders.

Of interest is the number of free parameters (i.e., weights) per output unit. This number gives a bound on the achievable reconstruction error. Table 6.2 shows the number of free parameters per output unit and the ratio of free parameters versus the number of weights (for calculating these numbers we have used only the weights from the input to the hidden units).

| structure | free parameters | $\frac{\text{free parameters}}{\#\text{weights}}$ |
|---|---|---|
| $2 \times 2/4$ | 1 | 1 |
| $3 \times 3/9$ | 1 | 1 |
| $3 \times 3/4$ | 1.44 | 4/9 |
| $4 \times 4/4$ | 4 | 4/9 |
| $5 \times 5/4$ | 4.59 | 9/25 |
| $2 \times 2/2$ | 2 | 2/5 |

Table 6.2: Number of Free Parameters Per Output Unit and Ratio of Free Parameters Per Total Number of Weights

From this analysis it is expected that the networks with more free parameters will achieve a lower reconstruction error (i.e., preserve more information). The ratio of free parameters versus the total number of weights tells us how costly the implementation is, the larger this number the better.

From Table 6.2, we see that the $4 \times 4/4$ pyramid has a large number of free parameters and at the same time a high ratio between free parameters and the total number of weights.

#### 6.3.3.1 PCA-Prism

The previous analysis indicated that we have to use more than one kernel in the pyramid. These kernels span the space of the first eigenvectors of the reduction window. When using different kernels in the pyramid, the resulting image will no longer be usable for image processing, because we are introducing an additional regular structure on the image. However, for other types of applications (e.g., image compression), this type of representation will still be useful (see section 6.5).

In order to use this pyramid also for other tasks, we can circumvent the problem by splitting the image into sub-images. For example, in the case of the $4 \times 4/4$ pyramid, we have four different types of cells, each with its own set of weights (see Figure 6.3). Therefore, we have four sub-images, each generated by one set of weights. Each such sub-image can be considered as a level of a $4 \times 4/16$ pyramid. We can now use the learning procedure on each of the sub-images independently. This will generate for each sub-image 4 additional images, resulting in 16 images at level 2 of the pyramid. The resulting pyramid structure is similar to a *prism* (Rosenfeld, 1985b) and depicted schematically in Figure 6.4. One can consider each branch of

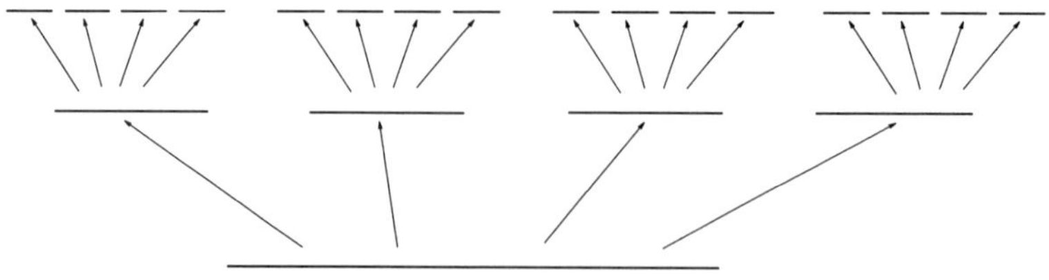

Figure 6.4: Structure of the $4 \times 4/4$ Prism

this prism as encoding a specific type of information of the original image.

In a prism, the total number of cells is kept constant at each level. In the case of the $4 \times 4/4$ pyramid, we can use each of the kernels learned as a generating kernel of a level of a usual $4 \times 4/4$ pyramid. This will result in the same number of cells at the next level, but now the information is split into different channels. Repeating this process keeps the total number of cells constant, but each branch of the pyramid, encoding a specific feature of the image, will have the structure of a $4 \times 4/4$ pyramid. The storage requirement for this type of pyramid is $\mathcal{O}(n^2 \log n)$ for an image of size $n \times n$, whereas for a conventional pyramid it is $\mathcal{O}(2n^2)$. This type of structure has the advantage that feature extraction is already performed during the construction phase of the pyramid.

## 6.4 Experiments

### 6.4.1 Oja Rule

We have simulated various pyramid structures with the Oja rule. The input images were first normalized to have zero mean. We have tested both a batch version (the weights are changed after presenting the whole image) and an online version (the weights are changed after every training example). The results for both procedures are almost identical except that the batch version is more stable but sometimes a little bit slower.

The results achieved were not very interesting. Most of the time, the kernel converged to identical weights. The convergence time is in the order of $5 - 10$ steps. Figure 6.5 shows 6 levels of a $3 \times 3/4$ pyramid generated by the Oja rule. Figure 6.6 shows four kernels of this pyramid. One can see that the kernels at the lower levels have a Gaussian shape, whereas at higher levels, different types of kernels develop. This is mainly due to the fact that at higher levels only a few pixels are used to train the kernel. These results are typical for most of the images we have tested. In general, the pyramids generated by the Oja rule are not very different from Gaussian pyramids. This is also true when local maxima detection

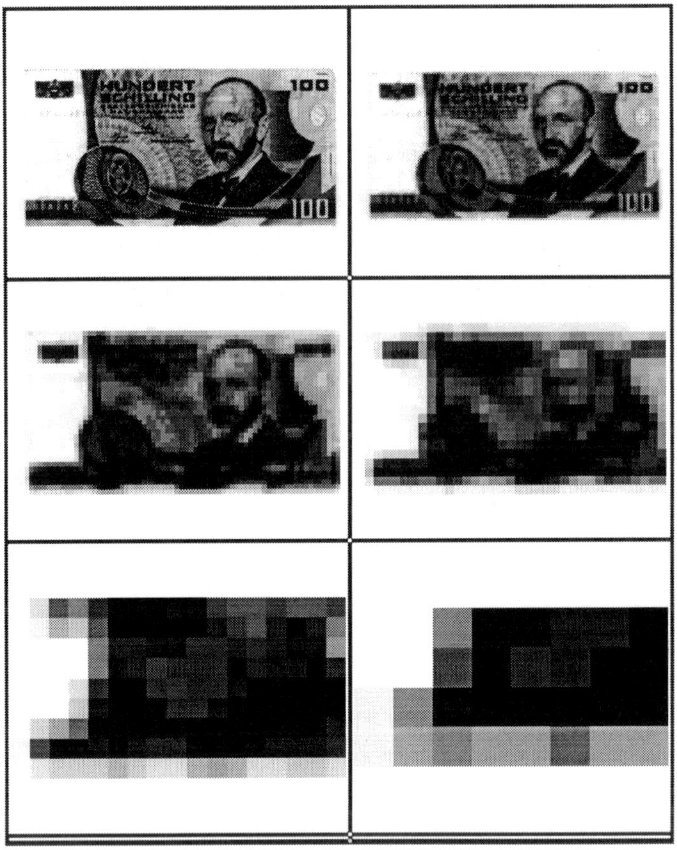



Figure 6.5: The first 6 levels of a $3 \times 3/4$ pyramid generated with the Oja rule

is performed, and only those cells are allowed to change their weights that are at local maxima position.

Only for special types of input images did the kernels deviate significantly from uniformity. For example, in Figure 6.7, one can see a kernel, that was generated for a $5 \times 5/4$ pyramid, where the input image was white on the left half and black on the right half. In this case, the kernel approximated the shape of the image.

Figure 6.6: The kernels of levels 1,2,5,8 (from top left to bottom right) generated by the Oja rule

Figure 6.7: Kernel generated by Oja rule for a test image

We have also tested the $n \times n \rightarrow 1 \rightarrow n \times n$ auto-associative back-propagation networks on these test images. The resulting kernels are almost the same as in the case of the Oja rule, but learning is more robust than for the Oja rule. In all our simulation runs, we achieved perfect convergence for a wide variety of parameter combinations. In the case of the Oja rule, we often had the effect that the learning algorithm did not converge due to a too large or too small learning rate. A further advantage of the auto-associative back-propagation network is that no explicit normalization of the input image is required. It is performed implicitly by the inclusion of a bias term in the hidden units. These results are in good agreement with the theory (Hornik & Kuan, 1992) that states that the multilayer versions of the principal component networks are more stable to train than the single layer networks.

### 6.4.2 $4 \times 4/4$ PCA-Pyramid

In order to evaluate the performance of the pyramids built with the auto-associative networks, we used the $4 \times 4/4$ pyramid. The only preprocessing we have done at the images is a linear contrast stretch in order to use the full input range of $[0 \ldots 255]$. All networks used in these experiments have been trained with a conjugate gradient algorithm, which converged in all our experiments. Figure 6.8 shows a typical learning curve for a network. During the first few steps the error decreases very rapidly (the first step is not shown in order to do a better scaling of the graph). Figure 6.9 depicts the weights of the hidden units during learning. After the first step, the weights are already close to their final value. After 10 steps, there are only small changes and minor improvements. Although it may take more than 100 steps until the learning algorithm has converged, the results after 10 steps of training are already acceptable. Therefore, a pyramid can be trained sufficiently fast. In the experiments reported here, we have used 100 steps for training each level.

Figure 6.10 depicts the original image and three levels of the $4 \times 4/4$ PCA-pyramid. The result is at the first sight disappointing. The regular structure imposed on the image is generated by the four different kernels of

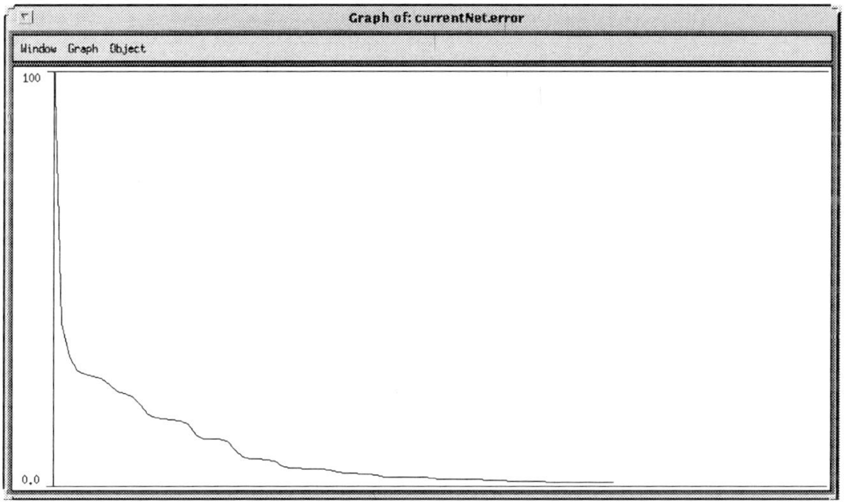


Figure 6.8: Learning curve for the auto-associative back-propagation network x-axis shows 100 learning steps, y-axis the squared error $\mathcal{E}$

the pyramid. It is clear that these images are not usable for image processing. Figure 6.11a shows level 1 of of the pyramid in Figure 6.10 split into four sub-images. The images are generated by the kernels A to D (see Figure 6.3a) from top left to bottom right. One can see that image D is considerably different from the other three images that look similar but have also differences. Figure 6.11b shows the same images as in Figure 6.11a but without sub-sampling.

From this type of representation, one can clearly see that different types of features are extracted in various sub-pyramids. Figure 6.12 shows the original image and two levels of the PCA-prism where the number of cells is kept constant during the pyramid construction process. For every sub-image of level 1, a new network was trained. Each branch of this structure is a $4 \times 4/4$ pyramid. The effects of sub-sampling are lowered by this type of representation, introducing additional redundancy. Therefore, this pyramid structure should be very robust against noise due to the redundant representation. Each branch of this structure encodes a specific feature of the original image.

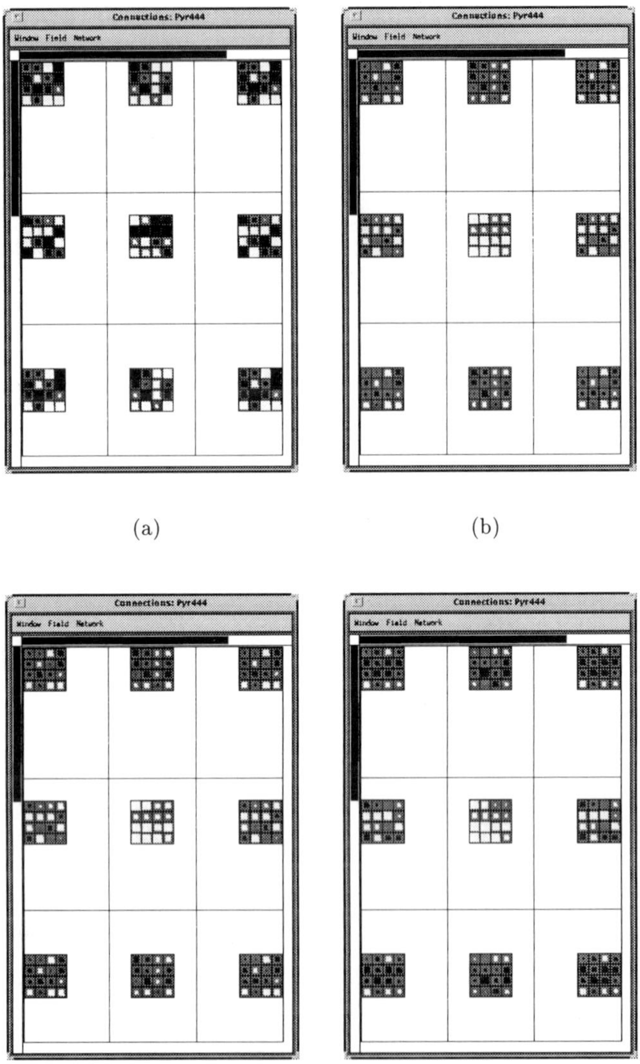


(a) (b)

(c) (d)

Figure 6.9: Weights of the hidden units of a $4 \times 4/4$ pyramid after 0 (a), 1 (b), 10 (c), and 100 (d) learning steps

Figure 6.10: First four levels of $4 \times 4/4$ PCA-pyramid

Figure 6.13a shows four levels of what might be called the PCA-Laplace pyramid. These images are generated similar to the levels of the Laplace pyramid. We take the difference between the level $L_i$ and the expanded level $L_{i+1}$. Figure 6.13b shows the corresponding levels of a $5 \times 5/4$ Laplace pyramid. One can see that the PCA-Laplace pyramid is much sparser, which means that the high resolution information can be much better reconstructed from the low resolution than in the case of the Laplace pyramid. This is not surprising because we train the pyramid in order to minimize this error. This important observation is used in section 6.5.

Another question we would like to demonstrate experimentally is what happens if we train a pyramid on one image and then apply this pyramid to another image without retraining. Figure 6.14 depicts the first level of the PCA-Laplace pyramid with the corresponding histogram for the 100ATS image. In the upper half of Figure 6.14, the pyramid was trained on this

(a) (b)

Figure 6.11: First level of PCA-pyramid (a) with sub-sampling, (b) without sub-sampling

image. In the lower half, the pyramid was trained on the house image. One can see from the histograms that the errors are only a little bit larger for the pyramid trained on the house image. With five additional steps of training on the 100ATS image, these pyramids are almost the same. From these results, we can conclude that we do not need to train the pyramid for each new image. And even if we want to train the pyramid, we can start from an already trained one, which will decrease the overall training time considerably.

From the results reported here, one can conclude that the PCA-pyramid is a valuable alternative multiresolution representation with many potential applications, one of which is discussed in the next section.

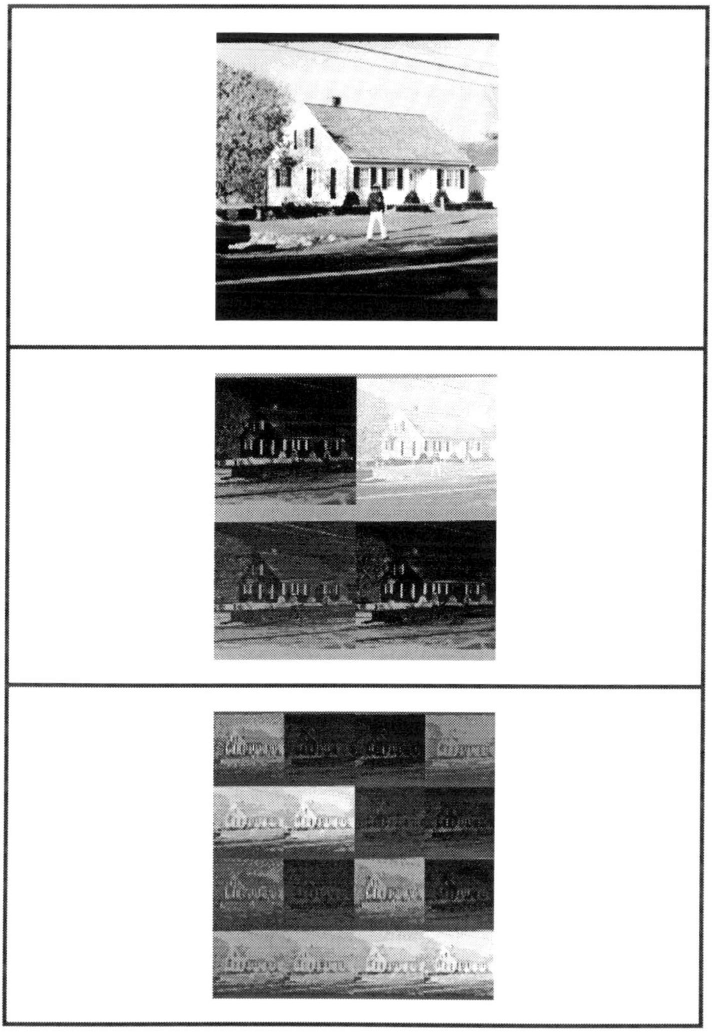

Figure 6.12: Levels 0, 1, and 2 of PCA-prism

(a) PCA-Laplace pyramid

(b) 5 × 5/4 Laplace pyramid

Figure 6.13: PCA-Laplace pyramid and 5 × 5/4 Laplace pyramid

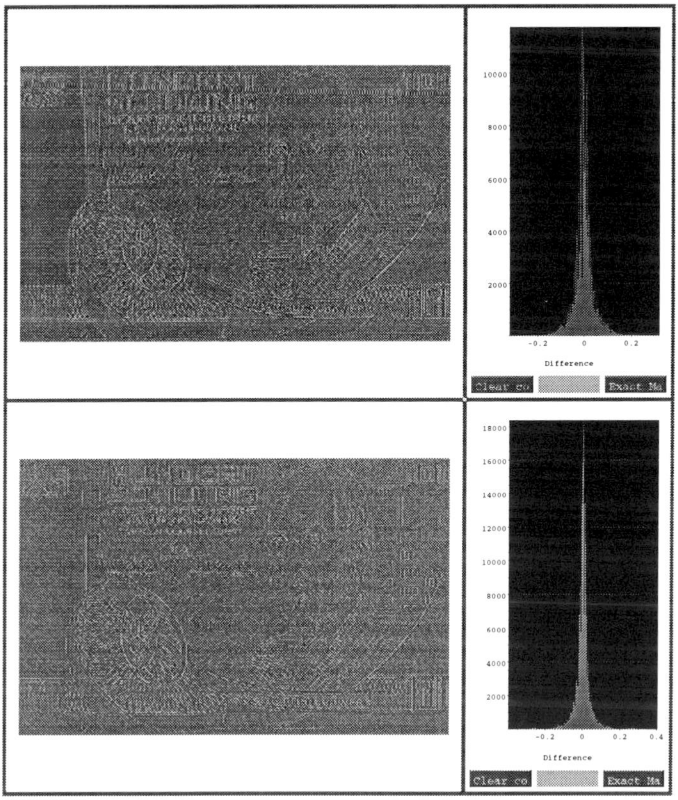


Figure 6.14: Reconstruction error of network trained on 100ATS image and trained on house image

## 6.5 PCA-Pyramid for Image Compression

In this section, we discuss the application of the PCA-Laplace pyramid to image compression. The method we propose is a combination between the Laplace pyramid (Burt & Adelson, 1983) and the auto-associative networks (Cottrell, Munro, & Zipser, 1987), which have both been used for image compression. The idea is similar to the Laplace pyramid; the only difference is that we learn the reduction and expansion function in order to minimize the squared Laplace level. Because this results in sparser Laplace levels (Figure 6.15),[4] we expect higher compression ratios. To build the PCA-Laplace pyramid, we construct the PCA-pyramid up to level $n$ (for a $512 \times 512$ image, $n$ is usually 4 or 5). After constructing the pyramid, we use the expansion functions to expand level $i$ in order to obtain an approximation of level $i-1$. Then we build the PCA-Laplace pyramid by subtracting level $i$ from the expanded level $i+1$; that is, $LP_i = L_i - E(L_{i+1}) = L_i - E(R(L_i))$. The original image can be completely recovered from level $n$ of the PCA-pyramid and the $n-1$ levels of the PCA-Laplace pyramid. Because the levels of the PCA-Laplace pyramid are extremely sparse (for example, see histogram in Figure 6.14), they can be compressed considerably by a conventional compression algorithm like the the Lempel–Ziv algorithm (Ziv & Lempl, 1977). In this case, we have a lossless (i.e., we do not lose any information) compression algorithm. In order to achieve higher compression ratios, we can additionally quantize the levels of the PCA-Laplace pyramid. For quantization, one could use either an adaptive method or a fixed quantization scheme. In Burt and Adelson (1983) and Mayer and Kropatsch (1989a, 1989b) different quantization methods for different levels of the pyramid have been used. At the low levels, only a few bits have been used, whereas at higher levels, many bits are used. In this case, the compression is lossy (i.e., loses information), because the original image cannot be exactly recovered. The compression ratio and the amount of loss can be controlled by the number of bits used to quantize the levels of the PCA-Laplacian.

[4]In the Laplace images, grey is 0, dark is negative values, and bright is positive values.

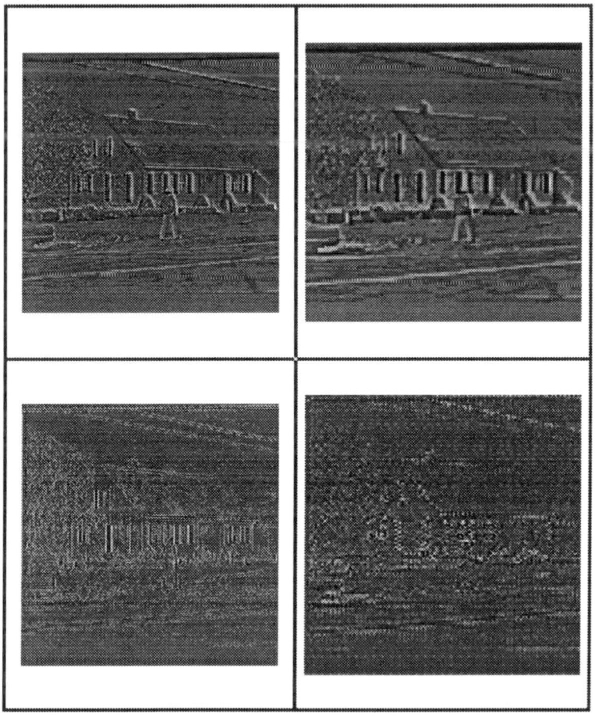

Figure 6.15: First two levels of a Laplace pyramid (upper half) and PCA-Laplace pyramid (lower half)

In order to measure the difference between the compressed and the original image, we have used, as in Cottrell et al. (1987) and Sanger (1989) the normalized mean squared error (NMSE), which is given by the mean squared error divided by the average squared intensity of the image.

$$NMSE = \frac{MSE}{< I^2 >} = \frac{< (I - C(I))^2 >}{< I^2 >} \tag{6.10}$$

where $I$ is the image and $C(I)$ is the compressed image. We measure the compression ratio by the amount of bits used to store $I$, divided by the amount of bits used to store $C(I)$.

| Quant. | Compression ratio | Bits/Pixel | NMSE |
|---|---|---|---|
| 3 Bit | 37.628 | 0.212 | 0.0172 |
| 4 Bit | 24.773 | 0.323 | 0.0019 |
| 7 Bit | 8.245 | 0.970 | 0.0000215 |
| no Quant. | 3.511 | 2.279 | 0.0 |
| Cottrell | 8.0 | 1.000 | 0.0059 |
| Sanger | 22.0 | 0.360 | 0.043 |
| $5 \times 5/4$ Laplace | 3.420 | 2.339 | 0.000087 |
| JPEG | 8.290 | 0.965 | 0.00139 |
| JPEG | 15.774 | 0.507 | 0.00348 |

Table 6.3: Compression Ratios and NMSE for Various Compression Methods

### 6.5.1 Experiments

The results reported here are obtained for the house image. For other images (e.g., the 100ATS image), we have achieved similar results. For the results reported here, we have used an uniform quantization which is fixed for all levels of the pyramid. As was shown in Burt and Adelson (1983) and Mayer and Kropatsch (1989b), the results can be improved by using a quantization that is lowered gradually from bottom to top.

Figure 6.16 shows the error images when the levels of the PCA-Laplacian pyramid are quantized with 3, 4, and 7 bits and the reconstructed image from the 7–bit Laplacian. Note that we have used the same lookup-table for the error images. To compress the levels of the PCA-Laplacian pyramid, we have used the standard UNIX compress program that implements a Lempel-Ziv algorithm.

From these images, one can see that the results with the 4– and 7–bit quantization are very good. One can visually perceive no difference between the reconstructed and the original image. Table 6.3 depicts the compression ratios and the NMSE on these images. These results compare favorably with the results in the literature (see Table 6.16). Cottrell et al. (1987)

Figure 6.16: Reconstruction error of house image with quantization of 3 bits, 4 bits, 7 bits, and reconstructed image

reported compression ratios of 8 (1 bit per pixel) with a NMSE of .0059 for their auto-associative network. Sanger (1989) reported a compression ratio of 22 (.36 bits per pixel) with a NMSE of .043 for the Generalized Hebbian algorithm. We have applied a $5 \times 5/4$ Laplace pyramid to the house image and achieved a compression ratio of 3.42 with a NMSE of .000087 for quantization with four bits of the Laplace levels. We have also included results achieved with JPEG.

It should be noted that the PCA-Laplace pyramid offers greater flexibility than the conventional Laplace pyramid, because it is trained on the image to compress and can therefore adapt to different images. But we do not need to train it on every new image we encounter, as can be seen from the results of Figure 6.14; even for completely different images the results are acceptable. For a typical application of image compression like teleconferencing, where the images share the same characteristics, training the pyramid once should be sufficient.[5]

We have tested what happens if we apply the same compression method to the levels of the PCA-Laplace pyramid. This results in a Laplacian of a Laplacian. If we could effectively compress the Laplace levels, we would obtain even higher compression ratios. However, as the experiment demonstrates (see Figure 6.17), the new Laplacian is almost similar to the original one. This result is clear, because due to sub-sampling we lose the high frequency information, which cannot be recovered. Therefore, this technique will not increase the compression ratio.

However, other methods we have not tested should improve the results. For example, other pyramid structures, such as the $5 \times 5/4$ pyramid, should yield better results because of the higher number of free parameters in the neural network (see Table 6.2). Another possibility to improve these results is to take the quantization into account during learning in the neural network. This is discussed in the next section.

### 6.5.2 Integration of Quantization

For the results reported in the previous section, we have used a fixed and uniform quantization scheme that can be improved by using adaptive quantizers like the Lloyd I algorithm, Kohonen's Feature Maps, learning vector quantization, or something similar. Such an approach was taken by Schweizer, Parladori, Sicranza, and Marsi (1991), who combined a Cottrell–type network with self–organizing feature maps. However, we can go further.

With the PCA–network, we minimize the squared Laplace level, which does not necessarily yield low compression errors. What we really want to minimize are the *quantized* Laplace levels. Usually, the Laplace levels have

[5]For moving images, other compression schemes also exist.

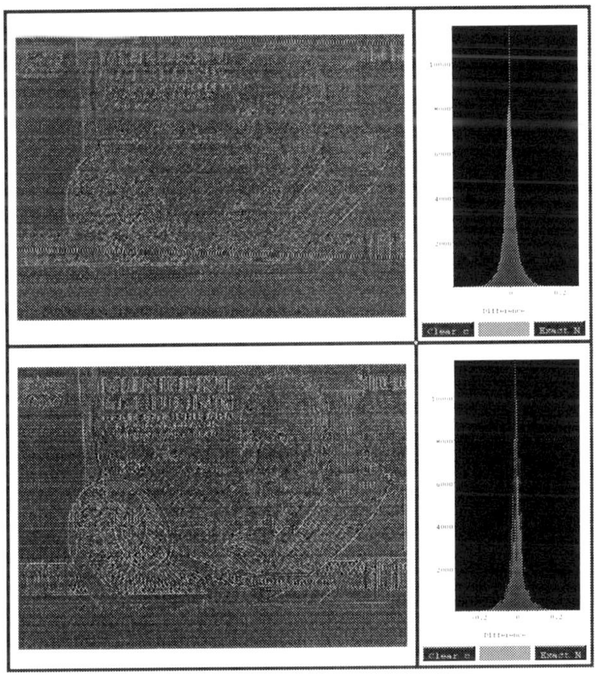

Figure 6.17: Laplace image of the Laplace image

an unimodally shaped histogram centered at zero. However, for the result of the compression (i.e., compression ratio and NMSE), it is irrelevant if we shift the histogram to the left or the right as long as we shift the quantization intervals in the same way. The best results could be achieved if we have a multimodal histogram with peaks centered at the quantization points.

Using neural networks for both PCA and quantization, this goal could be achieved by a modular network as shown in Figure 6.18 for the 1/2 pyramid. For quantization, we could either apply a vector quantizer to a whole patch of the Laplace level or use a scalar quantizer (as shown in Figure 6.18) for each pixel of the Laplace level. In the second case, we

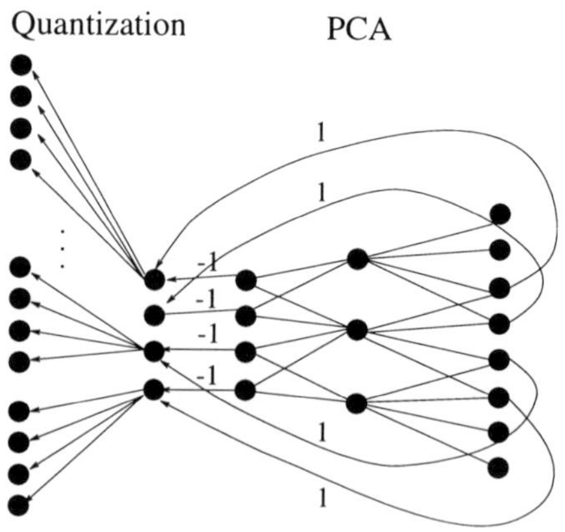


Figure 6.18: PCA-network and Quantization network

have to constrain the weights of the quantization network to be identical for every Laplace pixel. Because scalar quantization is simpler to analyze and uses less free parameters, we only consider this case.

As each quantization subnetwork can be treated separately (we only have to average the weight changes over all subnetworks), the following only considers the case of one output unit of the PCA–network.

The error to be minimized is the squared quantization error

$$E = \sum_p E_p \qquad\qquad E_p = \min_k ||c_k - l_p||^2,$$

where $p$ refers to the patterns in the training set, $c_k$ is the $k$th weight of the quantization network, and $l$ is the output of the PCA-Laplace unit.

Changing the weights of the quantization network by gradient descent leads to the LVQ1 rule of Kohonen

$$\Delta c_k = \begin{cases} 2\alpha(l_p - c_k), & \text{if } k = k_p \text{ is the winning unit,} \\ 0, & \text{otherwise.} \end{cases}$$

For the PCA-network, we can proceed similarly to back-propagation to obtain the rule

$$\Delta w_{ij} = -K\frac{\partial E_p}{\partial w_{ij}} = -K\frac{\partial E_p}{\partial l_p}\frac{\partial l_p}{\partial w_{ij}} = -K\frac{\partial E_p}{\partial l_p}\frac{\partial l_p}{\partial i'_p}\frac{\partial i'_p}{\partial w_{ij}} = -2K(l_p - c_k)\frac{\partial i'_p}{\partial w_{ij}}.$$

Of course, this is only one out of many possible algorithms. More elaborate minimization techniques than gradient descent could be used; similarly, LVQ1 could be replaced by a different quantization algorithm. But the basic idea of letting the quantization step and the the compression step adapt to each other remains unchanged.

## 6.6 Summary

In this chapter, we have introduced learning algorithms for pyramids. We have adapted and simulated several principal component algorithms for pyramid structures, yielding the so–called Principal Component Analysis Pyramids (PCA-pyramids). These algorithms can tune the reduction kernels of a pyramid to a specific image. The goal of the training algorithm is to preserve as much information as possible in the lower resolutions.

In the case of the Oja rule, we have found that most of the time the pyramid performs simple averaging. The interesting cases occur when different kernels are allowed in the pyramid construction. For this case, we have used three–layer auto-associative back-propagation networks to train the pyramid. We have derived constraints on the weights of the network depending on the structure of the pyramid. These pyramids can be used for image analysis, where feature extraction is already performed during the construction phase of the pyramid.

Another type of application we have dealt with is image compression. We have demonstrated that with the PCA-Laplace pyramid, we achieve better results than with the Laplace pyramid and with the methods proposed by Cottrell et al. (1987) and Sanger (1989). Due to the adaptive structure of the compression algorithm, it is applicable to a larger class of images. Moreover, we have shown how we can integrate the quantization step in our framework, yielding a modular neural network.

In the conclusion, we discuss another potential application of the PCA-pyramid to focus attention in hierarchical neural networks.

# Chapter 7

# Hierarchical Learning

In the previous chapter, we have seen how hierarchical structures can be trained by neural network algorithms. The resulting structures have interesting properties and are useful for various kinds of vision and image processing tasks. This chapter, too, is concerned with learning, however, from a different perspective. The question that heads this chapter is: "Can hierarchies help in learning?" The problem we would like to approach is this: Given a learning task (we confine ourself to supervised learning), what are the advantages gained by making use of hierarchical structures? Can we produce efficient learning algorithms? What are the properties of these algorithms?

The material presented in this chapter makes use of the work of Omohundro (Mel & Omohundro, 1991; Omohundro, 1989a, 1989b, 1990, 1991, 1992) and Landelius (1993). In particular, we consider tree–like structures to present our main ideas on hierarchical learning.

## 7.1 Visual Learning

As we have outlined in chapter 2, a visual system may be considered as a different number of maps $m_i$ $(1 \leq i \leq N)$

$$m_i : S_i \mapsto R_i$$

with $S_i$ the stimulus set and $R_i$ the set of responses.[1] To learn the map $m : S \mapsto R$ amounts then to learn the probability distribution $p(s, r)$ of the stimulus response pairs. There are at least two different ways of doing this:

1. Let the system model the average mapping from stimuli to responses and then apply some distribution with this function as its mean value. The parameters of the distribution may vary with the stimuli so that the system can be more or less certain on whether the mean value is appropriate or not.

2. Let the system estimate the distribution of stimulus response pairs in an analytic form. Because $p(r \mid s) = \frac{p(s,r)}{p(s)} \approx p(s, r)$, the conditional distribution can be calculated once the stimulus is known. Hence representing p(s,r), will allow for a response to be generated to the current stimulus $s_0$, by selecting a response at random from the distribution $p(s_0, r)$.

The implementation of both approaches outlined previously can be realized in two different ways:

1. Neural networks can be used to learn both, the average mapping from stimulus to response and for estimating the distribution $p(s, r)$. The first approach is done by multilayer perceptrons. The second approach falls in the domain of radial basis functions, which share close similarity to kernel regression methods as employed in statistics.

2. A currently not–so–popular way for learning of maps is to invent data structures that will store information about the distribution of stimulus response pairs in a way that will allow easy accessibility using the stimulus vector. The components of this vector will act as a key, or address, to the data structure. This could be considered as a memory structure much like the ones used in computers, and thus the problem of learning can be considered as a problem of memory

[1] In the following, we consider a single map and therefore skip the indices.

organization. Different data structures will decompose the stimulus response space in different ways producing different organizations of the memory. The goal is to develop structures that adapt to the underlying distribution so as to support fast access with little effort.

| | Neural Networks | Memory Structures |
|---|---|---|
| average mapping | Multilayer Perceptrons | Decision Trees |
| distribution estimation | Radial Basis Functions | Basis Function Trees |

Table 7.1: Different Possibilities and Prototypical Models Used to Learn Visual Maps

Table 7.1 summarizes the previous discussion and lists some prototypical models. A discussion of the neural network column has already been presented in chapter 2 and has dominated most parts of this book. In this chapter, we are concerned with the discussion on hierarchical memory structures; we discuss both models listed in the following sections. Before doing so, we discuss some differences between neural networks and hierarchical memory structures from a neural network point of view.[2]

- Standard neural networks maintain a complete model of its domain. The model is mostly wrong initially but gets better as more data becomes available. The net deals with all data in much the same way and has no representation for either strength of evidence behind a certain conclusion or for missing or uncertain input. The architecture is usually chosen before the data is presented, and the processing in the early phases resembles that in later ones. Human and animal learning, on the other hand, seem to proceed in a quite different manner (Omohundro, 1992). When the system has no or few experiences in a domain, every single experience is critical. Experiences are remembered more or less in detail in the early phases, and new responses

[2]This discussion follows the one in Landelius (1993).

are formed by generalizing from these stored experiences. Later on, when more data is available, more complex models can be formed, and there is no need for storing individual experiences. Instead the focus is on discovering regularities among them.

- The right topology of a network is hard to find; this problem is eliminated by hierarchical memory structures. In addition, an incremental refinement of the network topology is hard to do; this is very easily accomplished by hierarchical memory structures (e.g., addition of subtrees).

- In addition, global neural network models seem to be unnatural for learning. In such networks, all neurons must participate in the computation, even if only a few contribute to the output. The effort to process an input in a fully connected network capable of exact classification of $n$ memories is $\mathcal{O}(n^2)$. The units in the network form hyperplanes. The input is processed by letting each unit determine on which side of the hyperplane the input lies. This results in a waste of computational power, because many of these comparisons will be unnecessary. It is thus preferable to make use of a local model so that divide–and–conquer techniques can be applied, that is, every time new information is derived about the data, it is used to prune unnecessary further computations. Using hierarchical memories, the computational time can be reduced on the average to $\mathcal{O}(\log n)$.

## 7.2 Decision Trees

A decision tree classifies instances by recursively subdividing the feature space until each subdivided region contains instances of a single type. Each internal node has a function $f(\cdot)$ associated, which is used to split the feature space. Usually $f$ is a hyperplane dividing the feature space in two regions (i.e., each internal node checks if a data point $x = (x_1, x_2, \ldots x_n)^T$ is above ($f(x) > 0$) or below ($f(x) < 0$) the hyperplane). The resulting tree is a binary tree. Each terminal node (leaf) of a decision tree has a class label $\omega_i$ associated with it. Figure 7.1 shows a binary decision tree for a four–class problem.

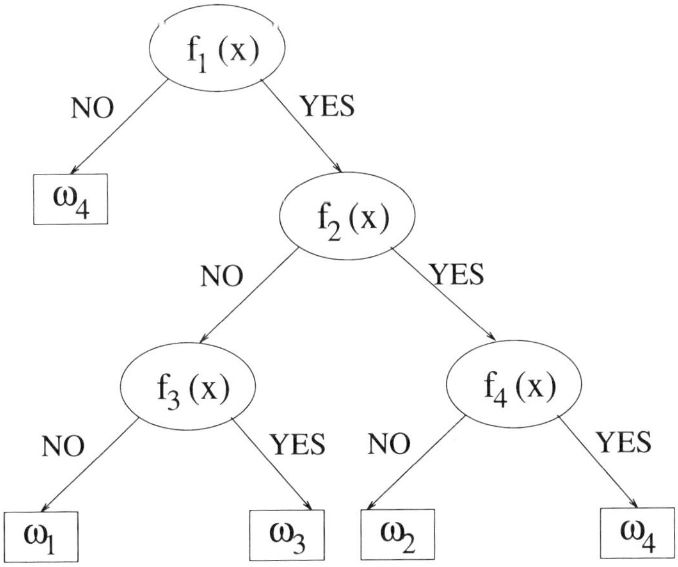


Figure 7.1: A decision tree for a four class problem

The best known decision tree algorithm is Quinlan's ID3 (Quinlan, 1986), which uses an information-theoretic measure to choose the function of the internal nodes. ID3-style decision trees use univariate functions in the internal nodes (i.e., $x_i - a > 0$, where $a$ is the value in the observed range of feature $x_i$). Because these tests are based on a single input variable, univariate trees can only divide the feature space orthogonally to a feature's axis (see also Figure 7.3). This may be inappropriate for problems with linearly related features (Breiman, Friedman, Olshen, & Stone, 1984).

Typically the construction of decision trees involves two steps:

1. Construct the decision tree until each training instance is correctly classified.

2. Prune the tree to avoid over-fitting of the training data.

Many different algorithms to construct decision trees have been proposed (e.g., Breiman et al., 1984; Friedman, 1973; Murthy, Kasif, & Salzberg, 1994; Quinlan, 1986; Utgoff & Brodley, 1990), which differ in the splitting and pruning criteria used as well as on the functions employed by the internal nodes (e.g., univariate or a linear combination of features; see also Figure 7.3).

It should be mentioned that we can convert any decision tree into an equivalent four-layer feed–forward network. This is demonstrated in Figure 7.2 for class $\omega_4$ of the decision tree in Figure 7.1. The idea in this construction

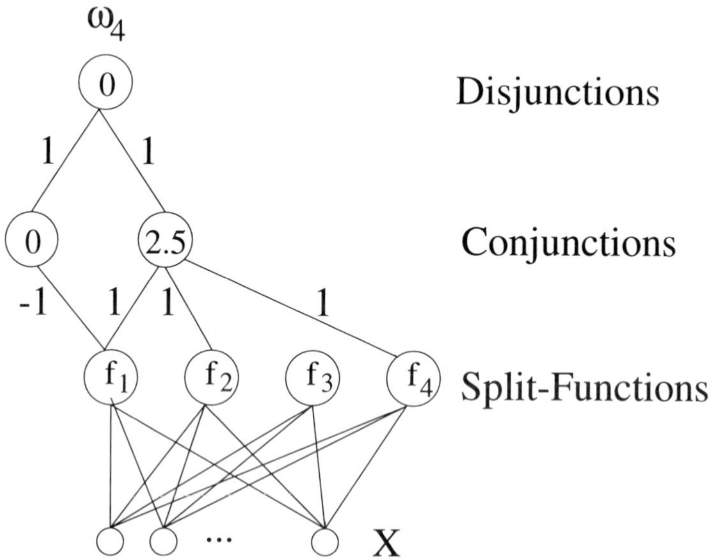


Figure 7.2: Equivalent feed-forward network for decision tree (numbers inside units denote thresholds)

is as follows: We can express any class as disjunctions of conjunctions of the functions $f_i$. For example, $\omega_4$ can be represented as $\omega_4 = \neg f_1 \vee (f_1 \wedge f_2 \wedge f_4)$, and this boolean expression can be easily converted into a three-layer network of threshold units. The equivalent network can be constructed as follows: The second layer computes the split functions $f_i$, the third layer computes the necessary conjunctions, and the fourth layer computes the disjunctions of the third layer.

This construction demonstrates also nicely one of the big advantages of decision trees over feed-forward networks. In order to classify a pattern in the feed-forward network, we have to evaluate all functions $f_i$, whereas in the case of the decision tree, we have to evaluate (on average) only a logarithmic number of these functions, which results in a considerable speed up in classification time.

### 7.2.1 Neural Tree Networks

Recently several authors have proposed a combination between neural networks and decision trees (e.g., Draper, Brodley & Utgoff, 1994; Guo & Gelfand, 1992; Utgoff & Brodley, 1990). The basic idea of these approaches is to use neural networks as functions in the internal nodes and to use neural network learning algorithms to learn these functions. Depending on the networks used as node functions, we have two different types of neural tree networks:

1. *Linear Networks:* Usually perceptrons are used as internal node functions; the resulting tree is binary and multivariate. Very efficient learning algorithms with guaranteed convergence properties exist to train the perceptrons. These trees can be generalized to so-called linear machine decision trees (Utgoff & Brodley, 1990). They generalize the two-category multivariate splits to n-category multivariate splits by the use of linear machines. At each internal node, a set of $R$ discriminant functions is trained to split the training set into $R$ classes. Special learning algorithms have been used in order to guarantee convergence even if the training set is nonlinear separable (this will happen at all internal nodes that do not have leaf nodes as their successors). Recently (Draper et al., 1994) showed how to use these trees for a focus of attention mechanism (cf. chapter 8) in a vision system; they have also demonstrated how to handle different costs of errors in such a network (e.g., when false positives are more expensive

that false negatives). In many applications, the superiority of neural tree networks over multilayer perceptrons has been demonstrated.

2. *Nonlinear Networks:* When replacing the two-layer perceptron by a multilayer perceptron, we can handle also nonlinear functions of the features. This results in trees with fewer levels as compared to linear trees. However, one should note that the complexity of the used networks has to be constrained in order to avoid the effect of having only one large network at the root node. A particular algorithm was proposed by Guo and Gelfand (1992). They used three-layer networks with one output unit to build a binary tree. A comparison to CART on a waveform and OCR classification task yields substantial improvement in classification accuracy, as well as a decrease in the number of nodes used and training time.

The work on neural tree networks demonstrates nicely how neural networks can be combined with hierarchical memory structures, with performance superior to both methods.

## 7.3 Basis Function Trees

Basis function trees can be considered as an extension of radial basis functions to hierarchical structures and can be used to represent the joint probability of stimuli and responses $p(s, r)$. The term basis function tree has been coined by Sanger (1991) and is used for function approximation methods with separable basis functions, which can be seen as polynomials of one–dimensional basis functions and can therefore be represented as a tree structure.

We use the term basis function tree more rigidly than Sanger and consider only trees with radial symmetric basis functions in their internal nodes. Most of the discussion is confined to Gaussian functions (this makes also the usage of these trees as representing probability distributions possible).

The first question to ask is which tree structure to use. This corresponds to a partitioning of the input space (signal space). Figure 7.3 shows several different possibilities of such tree structures.[3]

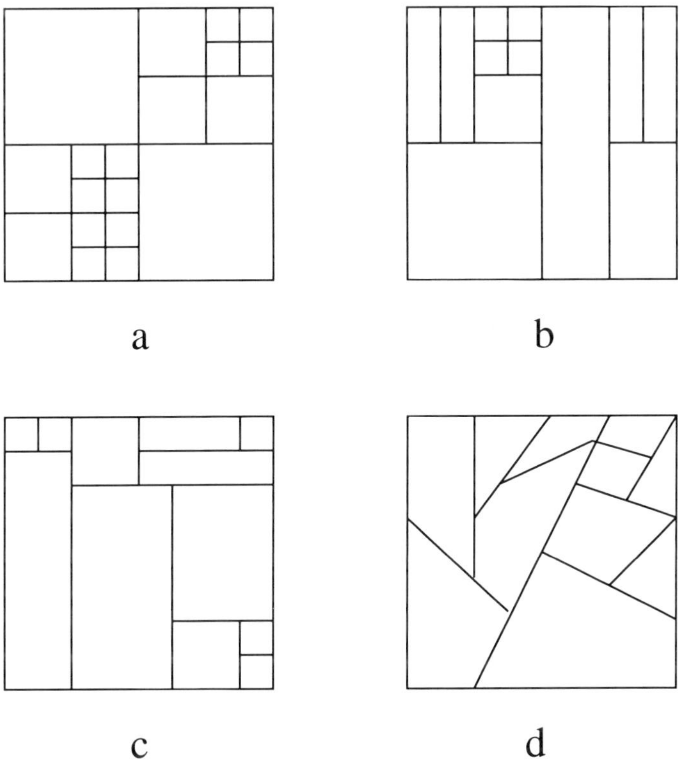


Figure 7.3: Different tree structures in a two-dimensional input space. The partitions are represented by the leaf regions of each structure.

In a multilevel grid, a fixed-sized grid is used to partition the space at different levels (see Figure 7.3a). First, the whole space is decomposed coarsely by the grid. This partitioning then continues recursively with each partition until the space is covered densely enough. Well-known structures of this type are quadtree and the octtree (Samet, 1990), which arise when

[3]adapted from Landelius, 1993

two- and three-dimensional spaces are decomposed with grids of size $2 \times 2$ and $2 \times 2 \times 2$, respectively. This structure is also closely related to regular image pyramids (i.e., it is a substructure of a regular pyramid). In higher dimensional spaces, this structure tends to use large amounts of memory for a given accuracy of representation.

Modifying the multilevel grid and allowing splits only along one dimension at a time results in a $k - d$ trie[4] (Figure 7.3b). The $k$ dimensions are still split in half, but each node in the trie now contains information about which dimension $d$ to split.

A natural relaxation of the $k - d$ trie is to allow for splits at any position along a dimension, not just at the midpoint; this results in the so–called $k - d$ tree. This is illustrated in Figure 7.3c. The $k-d$ tree is an ordinary binary tree, but each node contains information both about which dimension to split and also where along that dimension to position the split. The same structure results in univariate decision trees. Each node in a $k - d$ tree corresponds to a hyper-rectangular piece of space, in which some or all dimensions may be infinite. This holds also true for the leaves of the tree.

The last structure is the $k - \mathbf{e} - d$ tree. The partitions in this structure no longer have to be along the dimensions but can be along any direction in space (Figure 7.3d). This means that the tree now has to store information about the direction $\mathbf{e}$ of the split and also where along this direction the split is made. The vector $\mathbf{e}$ can be seen as the normal vector of the hyperplane that partitions the space at a node. This structure results in multivariate decision trees and neural tree networks with perceptrons in their internal nodes. It has also some similarity to irregular pyramids.

Omohundro (1989a) compared five algorithms for $k - d$ tree construction (four top–down algorithms including two online insertion algorithms and one bottom–up algorithm). The bottom–up algorithm has done an excellent job in finding good tree structures, however, it has a high computational cost. Any of these algorithms can be used for our purpose.

In order to represent probability distributions ($p(s, r)$ in our case), Omo-

[4]note this is not a misspelling

hundro has introduced the so-called *bump trees* (Omohundro, 1991). Each leaf of the tree corresponds to kernel function (e.g., Gaussian[5]). There are also functions associated with each internal node with the constraint that a function in an interior node must be everywhere larger than the functions of its children (see Figure 7.4). This is especially important because

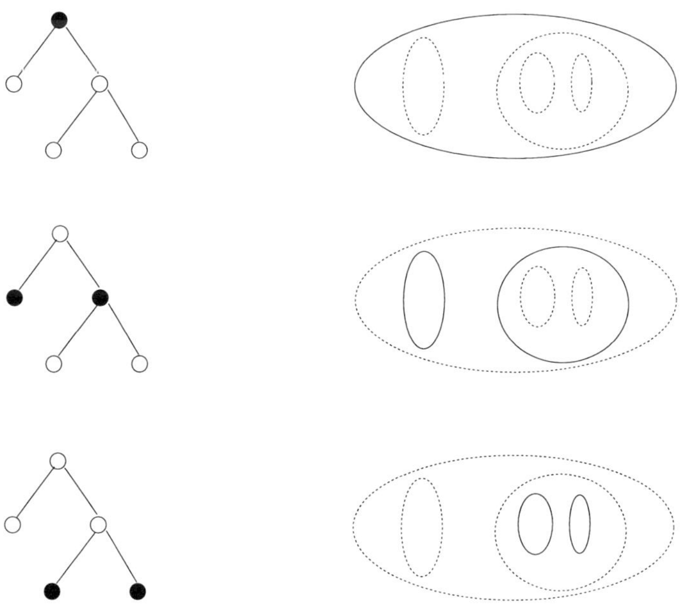

Figure 7.4: The bump functions of a bump tree

branch-and-bound techniques can then be used to answer question about the data in an efficient way (i.e., in $\mathcal{O}(\log n)$ steps). The simplest kind of query presents a point in space and asks for all leaf functions that have a value at that point that is larger than some predefined value. Another typical example in the case of Gaussians is to request all Gaussians in the set whose value at a specified point is within a specified fraction of the Gaussian whose value is largest at that point.

[5]Any probability density function, with a finite number of discontinuities, can be approximated arbitrarily close by a sum of normal distributions (Sorenson & Alspach, 1971).

Omohundro (1991) compared bump trees with radial basis function neural networks on a robot arm mapping task. The error is significantly lower for bump trees than for Gaussian radial basis function neural networks, although the retrieval time is considerably faster. Landelius (1993) demonstrated how bump trees can be used for reinforcement learning and representing "Behavior distributions." These results demonstrate that bump trees and similar structures are an efficient alternative to radial basis functions and should be considered in more detail in the future.

## 7.4 Evaluation

In this section, we discuss several properties of hierarchical memory structures from the view point of the problems that we have listed in section 2.1.4.1. That is:

1. Global versus local models
2. Curse of dimensionality
3. Bias/variance dilemma

**Global Versus Local Models:** The use of hierarchical organizations implies the use of local models. Hierarchies get their efficiency by divide–and–conquer techniques, that is, global problems are decomposed into local ones. Therefore, any model to be used with a hierarchy has to be a local one. This can be done in two ways either by being a priori local (e.g., bump functions) or by restricting it to a local domain (e.g., perceptrons in neural tree networks).

**Curse of Dimensionality:** The divide–and–conquer approach is also efficient to fight against the curse of dimensionality. From Figure 7.3, one can see that hierarchical structures can adapt to the underlying distribution. Therefore, resources are only used in those parts of the space where there is sufficient data available. In parts with no or only few samples, the approximation is very coarse.

**Bias/Variance Dilemma:** Using the structures described, we introduce bias. Particularly the bias we introduce is at least one of the following types (see also the discussion in section 2.1.4.1):

1. Locality
2. Smoothness

Omohundro (1992) argued that these kinds of biases are harmless ones, and indeed many vision researchers make implicit or explicit use of them. By introducing bias in the system, we decrease the variance, and therefore, we can expect good results from learning (Geman et al., 1992). However, one should always be cautious in applying a biased learning system to a new domain; it has to be checked if the bias introduced is indeed harmless in that domain.

In the visual learning section (section 7.1), we have described that human and animal learning operate in different modes. First, every sample is remembered more or less in detail (this might be considered as a simple nearest neighbor classification), and when more examples are available, more complex models can be formed (i.e., we begin building our hierarchy by fitting parameters to models). This suggests a bottom-up construction of the hierarchy. However, most of the models presented in this chapter proceed top-down.

An approach for constructing flexible hierarchies in a bottom-up fashion is the irregular pyramid as presented in chapter 3. In order to apply the irregular pyramid to learning tasks, we need to extend it to higher dimensional spaces and incorporate some kind of bump-tree functionality into this structure. Another major long term-goal is to develop a family of classifiers that has on the one end, a nearest neighbor classification and moves along to k-nearest neighbor methods to bump-tree structures, which would resemble very closely the previously outlined learning approach.

## 7.5 Summary

This chapter has considered the use of hierarchical structures in learning. We have presented several different approaches like decision trees, neural tree networks, basis function trees and bump trees. The basic idea behind these models is always the same: The input space is recursively and adaptively divided into several regions. We have shown that these models can be either used for classification or representing probability distributions. We have also discussed how these models can make some of the fundamental problems of learning more tractable. Several cited experimental studies have demonstrated the superiority of this approach over conventional methods and standard neural network models. We have outlined several possibilities for further research, that is, the incorporation of irregular pyramids and bottom-up construction algorithms, as well as the considerations of families of learning algorithms.

# Chapter 8

# Conclusion and Outlook

Since the mid-1980s there are two different areas of research in the pattern recognition and computer vision community. The one is more oriented towards neural network models; the other is oriented to the classical fields of statistical pattern recognition, knowledge–based methods, mathematical modeling, and so forth. The advocates of these two directions have often described them as different approaches. One of the goals of this book is to transfer knowledge between these two fields. The goal of all these approaches is the same: "Let a computer 'see' as we do." What is different is the vocabulary of description. So the question is not: "Do neural networks provide anything new?" The right way to ask is: "Do we gain new insights or more general solutions by a description in terms of neural networks?"

Throughout this book, we have seen that it is sometimes harder to describe an algorithm in terms of neural networks, but the results obtained offer new ways of solutions, or they are more efficient. Let us mention three examples:

1. We have described the $2 \times 2/2$ curve pyramid in terms of neural networks. This description has allowed us to generalize the curve pyramid also to processing of curve strength. From this formulation, it was easy to see that it can also be described in terms of fuzzy logic.
2. In order to describe the maximum pyramid in terms of neural networks, we had to use a special type of a Winner Takes All network.

To find a maximum of numbers is an easy task for computers. We just look at every number and see which one is the largest. However, the neural network solution has the advantage that its convergence time is a constant, irrespective of how many numbers are input to our algorithm.

3. We have described how to combine the transformation and the quantization step for image compression. This was only possible by description of these two parts as a neural network. With this common description, it was possible to develop a common learning algorithm.

Formulating an algorithm in terms of neural networks, we are forced to use a very elementary vocabulary, that is, units, connections, weights, update functions, and so forth. However, this fine–grained description offers more flexibility, and the resulting algorithms are often more general than their conventional counterparts. Neural network research is still in its infancy, therefore, we do not have many building blocks available to describe larger systems. But it should be one of the goals of neural network research to take existing solutions and combine them to larger systems. In this book, we have offered some of the building blocks for computer vision. Let us briefly evaluate what we have achieved and what are the open points for further research in that direction.

## 8.1 Contributions and Evaluation

One of the main points in this book was an in-depth comparison of neural networks and image pyramids. We have shown that for any image pyramid, we can build an equivalent neural network. To our knowledge, such an analysis has not been done before. Out of this comparison, several new algorithms and enhancements of existing algorithms have arisen. Let us mention the main contributions in this respect:

- Neural network classification in a hierarchical structure;
- Shift variance proof of hierarchical neural networks;

- Hopfield neural networks for calculating maximum independent sets of a graph;
- Segmentation by irregular pyramids controlled by Hopfield networks;
- Distributed representation of curve relations;
- Sigmoid pyramids;
- Neural curve pyramid ⇒ fuzzy curve pyramid;
- Learning algorithms for pyramids ⇒ PCA-pyramid;
- Neural Laplace pyramid for image compression;
- Usage of hierarchical structures in learning.

Because we have studied so many different things, we have not evaluated all of them in full detail. Open points for further research are:

- Study of the classification algorithms in other types of pyramid structures;
- Incorporation of edge information in the Hopfield decimation process;
- Theoretical investigation of setting the weights of the Hopfield network for adaptive decimation;
- Theoretical investigation of properties of the Sigmoid pyramid and adaptive selection of the steepness parameter $\beta$;
- Neural curve pyramid for distributed representations of curve relations;
- Implementation and experiments with the fuzzy curve pyramid;
- Evaluation of learning algorithms for pyramids for a broader class of images;
- Usage of irregular pyramids to build hierarchical memory structures;

Many things remain to be done in order to evaluate and enhance the proposed algorithms. But besides these short term–goals, we also study new problems within our general methodology of pyramidal neural networks. One of the most demanding and interesting ones is an attention mechanism for vision. The outlook discusses some general ideas on this topic.

## 8.2 Outlook

Vision is an inherently complex process, especially when dealing with complex images (e.g., outdoor scenes) in a dynamic environment, where decisions have to be taken in real time (e.g., navigation of a robot). The work of Tsotsos (1990, 1991) shows that a hierarchical structure is necessary to cope with complexity. But a hierarchical structure alone is not sufficient. Throughout this book, we have seen several examples of hierarchically structured neural networks. We have also seen in section 3.1, that it is sometimes advantageous to have access to parts of the image in full resolution. This means that the system must have the possibility to *focus its attention* on specific parts of the image. Attentive mechanisms appear also to be at the very basis of biological vision systems. There are two different types of attention, *external* (overt) and *internal* (covert). External means that there must be an active sensor that can decide where to look next (e.g., eye movements). The result of a shift of attention is a new image. In the case of internal attention, we have to decide which parts of a given image to process. Both types of attention mechanisms are present in human vision. External attention is associated with the nonuniform resolution of the human retina. Saccades are used to bring interesting regions in the high resolution part of the retina. But surprisingly also internal shifts of attention take place in human vision (see, for example, the experiments by Moran & Desimone, 1985 and Desimone & Schein, 1987). External shifts of attention are usually more costly in terms of time and energy consumption than internal shifts of attention. It is still an open question if the same mechanisms underlie these two forms of attention.

A lot of different architectures have been proposed that allow a system to focus its attention (e.g., Ahmad, 1991; Fukushima, 1986; Milanese, 1992; Olshausen, Anderson, & Van Essen, 1992; Tsotsos, 1991; see Milanese, 1990, 1994 for an overview). Most of these models specify only the architecture that allows a system to focus its attention on specific parts of an image, but they do not address the question of where to focus attention. Most of them speak simply about change, motion, contrast, and so forth, but do not give a general answer to the question of where to focus attention. In order to answer this question, one has to answer first the question, "What is the purpose of focusing attention?" We have already given a partial answer, that is, in order to reduce the complexity of vision, only "important" parts of the image are processed. But what is important for a vision system? One may say this depends on the task to perform (e.g., obstacle avoidance, identifying food, etc.). But one can answer this question more generally without a reference to the task to perform. What is important is what we do not know yet, what is new, and what we cannot predict. For example, sitting in our office, we know where the desk is, where the chairs are, and so forth; we do not need to focus our attention on these things. But if suddenly the lamp falls from the ceiling, this is new, and we cannot predict it; therefore, we focus our attention on it.

Consider a computer vision system in a static environment. Because nothing moves (also the system does not move), there would be no need for this system to have a focus of attention everything is predictable (i.e., it is the same as it was before). So having once interpreted the scene, it knows everything about it and can turn off the camera. In a dynamic environment, so many things may happen, therefore, it is important to constantly watch it. But due to processing limitations, we can not watch everything simultaneously, therefore, we focus our attention at those objects we cannot predict.

So our theory is that attention is closely coupled with predictions about the environment. If the prediction we make is correct, there is no need for further processing; however, if the prediction is wrong, there is evidence that something important is going on.

Seeing attention in this way allows us also to incorporate learning. The predictions about the environment are learned. So whenever we have a mismatch between the prediction and the real environment, we can do two things: change the prediction (and processing) mechanism and focus attention on those parts of the image where the prediction was wrong. It should be noted that the predictions are not necessarily conscious, most of them will be implicit.

Considering learning in the framework of predictions avoids also the problem of having a separate teacher signal for every neuron. The teacher is simply the environment. Another point worth mentioning is that this method can also be applied inside a network, where one module predicts the outcome of another module. There has been some work in that direction (e.g., Becker, 1992; Becker & Hinton, 1992; Schmidhuber & Prelinger, 1993).

For the case of external attention, we have already seen that the goal of a shift of attention is to bring interesting regions to the high resolution part of the retina. Assuming a hierarchical structured vision system (see also the discussion in the introduction), we can state the same goal for internal attention, that is, to access the high resolution parts of the image via the resolution hierarchy.

The models proposed by Tsotsos (1991) and Olshausen et al. (1992) specify a concrete biologically plausible mechanism of how this might be accomplished. The main difference between these models is how the high resolution part is routed through the resolution hierarchy.

Let us now translate these general ideas to hierarchical neural networks. In order to simplify the discussion, we assume that we have only one hierarchical structure (i.e., only one type of feature), and we are dealing with a mechanism for focusing internal attention. Figure 8.1 depicts the general situation we are dealing with. We have a resolution hierarchy that is constructed via the reduction functions $R_i$. For every reduction function, we have the corresponding expansion (prediction) function $P_i$. In order to measure the dissimilarity between the prediction and the original, we have a distance function $d$. The result of application of the distance function $d(I_i, P_i(R_i(I_i))) = E_i$ is an "error map" with the same resolution as $I_i$.

Different types of functions are possible for reduction, prediction, and di-

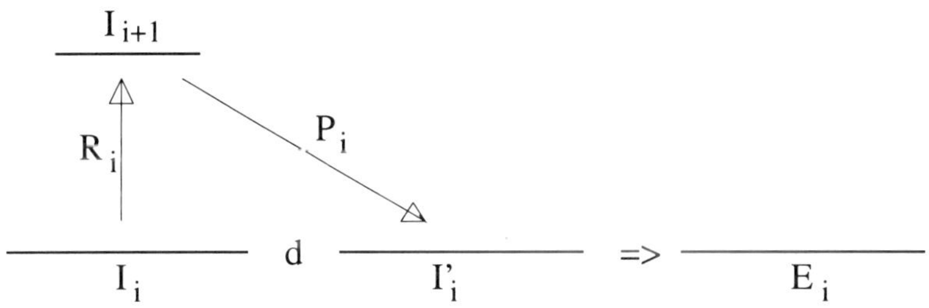


Figure 8.1: Attention in hierarchical neural networks

stance, but in order to be computationally efficient, we require them to be local and independent of the sequence of calculation; therefore, a parallel implementation is possible.

With this scheme we can now define a mechanism for focusing attention. We first construct the resolution hierarchy with the reduction functions $R_i$, then we predict from the lower resolutions the high resolution with the prediction functions $P_i$ and measure the error with the distance function $d$. If the error is above a threshold, we have found a candidate region to focus our attention on. At the same time, we can change the reduction and prediction functions with a learning algorithm. This can be done in a similar manner as was discussed in chapter 6. One should note that we have used learning algorithms that preserve as much information as possible, given the constraints of the architecture. Therefore, the shifts of attention are minimized, which in turn saves time to interpret the scene.

Those regions we have identified as candidate points for focusing attention on, can be ordered according to the prediction error, and those with the largest error are examined first. These regions are then routed via the resolution hierarchy through the network, in order to allow an examination in full resolution. This can be accomplished with mechanisms as proposed by Tsotsos (1991) or Olshausen et al. (1992). When these regions have been examined, they are inhibited, so that other regions with less error in the prediction can also be examined.

This model allows us to focus attention in hierarchical neural networks. The description we have presented here is very general and several things need further refinement. However, the general ideas that attention is coupled with predictions, and predictions are coupled with learning seem to be very promising. First implementations for simple cases (Bischof & Hraby, 1993) show promising results. Future results will hopefully demonstrate the strength of this approach.

# Bibliography

Abrahamson, K., Dadoun, N., Kirkpatrick, D. G., & Przytycka, T. (1987). *A simple parallel tree contraction algorithm.* Tech. Rep. No, 87–30, Vancouver: University of British Columbia, Department of Computer Science.

Ackley, D. H., Hinton, G. E., & Sejnowski, T. J. (1985). A learning algorithm for Boltzmann machines. *Cognitive Science, 9*, 147–169.

Ahmad, S. (1991). *VISIT: An efficient computational model for human visual attention.* Tech. Rep. No, 91–049, Berkley, CA: International Computer Science Institute.

Albrecht, R. F., Reeves, C. R., & Steele, N. C. (Eds.) (1993). *Artificial neural networks and genetic algorithms.* Vienna: Springer–Verlag.

Almeida, L. B. (1988). Backpropagation in perceptrons with feedback. In: *Neural Computers*, Eckmiller, R. & van der Malsburg, C. (Eds.), (pp. 199–208 ). Berlin: Springer-Verlag.

Anderson, J. A., Pellionisz, A., & Rosenfeld, E. (Eds.) (1990). *Neurocomputing II: Directions of research.* Cambridge MA: MIT Press.

Anderson, J. A. & Rosenfeld, E. (Eds.) (1988). *Neurocomputing: Foundations of research.* Cambridge MA: MIT Press.

Anderson, T. W. (1958). *An introduction to multivariate statistical analysis.* New York: Wiley.

Anshelerich, V. V., Amirikian, A. V., Lukashin, A. V., & Frank-Komenetskii, M. D. (1989). On the ability of neural networks to perform generalization by induction. *Biological Cybernetics, 61*, 125–128.

Baldi, P. & Hornik, K. (1989). Neural networks and principal component analysis: Learning from examples without local minima. *Neural Networks, 2*, 53–58.

Baldi, P. & Hornik, K. (in press). Learning in linear neural networks: A survey. *IEEE Transactions on Neural Networks*.

Ballard, D. H. (1987). Interpolation coding: A representation for numbers in neural models. *Biological Cybernetics*, *57*, 389–402.

Barnard, E. & Casasent, D. (1990). Shift invariance and the Neocognitron. *Neural Networks*, *3*, 403–410.

Barron, A. (1992). Neural net approximation. In: *Proceedings of the 7th Yale Workshop on Adaptive Learning Systems*, (pp. 69–72 ). New Haven, CT: Yale University Press.

Barto, A. G. & Anandan, P. (1985). Pattern recognizing stochastic learning automata. *IEEE Transactions on System, Man and Cybernetics*, *15*, 360–375.

Baum, E. B. & Haussler, D. (1990). What size net gives valid generalization? *Neural Computation*, *1*, 151–160.

Becker, S. (1992). *An information-theoretic unsupervised learning algorithm for neural networks*. Unpublished doctoral dissertation, University of Toronto.

Becker, S. (1993). Learning to categorize objects using temporal coherence. In: *Advances in Neural Information Processing Systems 5*, Hanson, S. J., Cowan, J. D., & Giles, C. L. (Eds.), (pp. 361–368 ). San Mateo, CA: Morgan Kaufmann.

Becker, S. & Hinton, G. E. (1992). Self-organizing neural networks that discover surfaces in random-dot stereograms. *Nature*, *335*, 161–163.

Bederson, B. B. (1992). *A miniature space-variant active vision system: Cortex I*. Unpublished doctoral dissertation, New York University, Courant Institute.

Bertin, E. & Bischof, H. (1993). *Voronoi pyramids controlled by Hopfield networks*. Tech. Rep. No, PRIP-TR-24, Technical University Vienna, Dept. for Pattern Recognition and Image Processing.

Bertin, E., Parazza, F., & Chassery, J. M. (1993). Segmentation and measurement based on Voronoi diagram: Application to confocal microscopy. *Computerized Medical Imaging and Graphics*, *17*(3), 175–182.

Betzdek, J. C. (1981). *Pattern recognition with fuzzy objective function algorithms*. New York: Plenum Press.

Bilbro, G., Mann, R., Miller, T. K., Snyder, W. E., Van den Bout, D. E., & White, M. (1989). Optimization by mean field annealing. In: *Advances in Neural Information Processing Systems*, Touretzky, D. S. (Eds.), volume I, (pp. 91–98 ). San Mateo, CA: Morgan Kaufmann.

Bischof, H. & Hraby, K. (1993). Focusing attention in hierarchical neural networks. In: *Substance Identification Analytics*, Flanagan, J. L., Mammone, R. J., Brandstein, A. E., Pike, E. R., Thomopoulos, S. C. A., Boyer, M.-P., Huang, H. K., & Ratib, O. M. (Eds.), volume 2093 of *Europto Series*, (pp. 230–241 ). Bellingham, WA: SPIE The International Society for Optical Engineering.

Bischof, H. & Pinz, A. (1990). Verwendung von neuralen Netzwerken zur Bestimmung der Baumart aus digitalen Rasterbildern (Usage of neural networks for determination of tree species from digital images). In: *Wissenbasierte Mustererkennung*, Axel, P. (Eds.), OCG-Schriftenreihe 49, (pp. 149–161 ). Wien: Oldenbourg–Verlag.

Bischof, H. & Pinz, A. (1991). Visualization methods for neural networks. In: *Konnektionismus: Beiträge aus Theorie und Praxis (Connectionism: Contributions of Theory and Practise)*, Ziegeler, H. G. (Eds.), volume 3, (pp. 1–14 ). Schriftenreihe der ÖGAI-Austrian Society for Artificial Intelligence.

Bischof, H. & Pinz, A. (1992). The invariance problem for hierarchical neural networks. In: *Neural and Stochastic Methods in Image and Signal Processing*, Su-Shing, C. (Eds.), volume 1766, (pp. 118–129 ). Bellingham, WA: SPIE-The International Society for Optical Engineering.

Bischof, H., Pinz, A., & Kropatsch, W. (1992a). Visualization methods for neural networks. In: *Proceedings of the 11.International Conference on Pattern Recognition*, volume II, (pp. 581–585 ). Los Alamitos, CA: IEEE Computer Society.

Bischof, H., Schneider, W., & Pinz, A. (1992b). Multispectral classification of landsat-images using neural networks. *IEEE Transactions on Geoscience and Remote Sensing*, *30*(3), 482–490.

Bister, M., Cornelis, J., & Rosenfeld, A. (1990). A critical view of pyramid segmentation algorithms. *Pattern Recognition Letters*, *11*(9), 605–617.

Blanford, R. P. & Tanimoto, S. L. (1988). Bright spot detection in pyramids. *Computer Vision, Graphics, and Image Processing*, *43*(2), 133–149.

Bourlard, H. & Kamp, Y. (1988). Auto-association by multilayer perceptrons and singular value decomposition. *Biological Cybernetics*, *59*, 291–294.

Breiman, L., Friedman, J., Olshen, R., & Stone, C. (Eds.) (1984). *Classification and regression trees*. Belmont, CA: Wadsworth.

Burt, P. J. & Adelson, E. H. (1983). The Laplacian pyramid as a compact image code. *IEEE Transactions on Communications*, *31*(4), 532–540.

Carpenter, G. A. (1989). Neural network models for pattern recognition and associative memory. *Neural Networks*, *2*, 243–257.

Carpenter, G. A. & Grossberg, S. (1987a). ART2: Self-organization of stable category recognition codes for analog input patterns. *Applied Optics*, *26*, 4919–4930.

Carpenter, G. A. & Grossberg, S. (1987b). A massively parallel architecture for a self-organizing neural pattern recognition machine. *Computer Vision, Graphics and Image Processing*, *37*, 54–115.

Carpenter, G. A. & Grossberg, S. (1990). ART 3: hierarchical search using chemical transmitters in self-organizing pattern recognition architectures. *Neural Networks*, *3*, 129–152.

Carpenter, G. A., Grossberg, S., & Reynolds, J. H. (1991a). ARTMAP: Supervised real-time learning and classification of nonstationary data by a self-organizing neural network. *Neural Networks*, *4*(5), 565–588.

Carpenter, G. A., Grossberg, S., & Rosen, D. B. (1991b). Fuzzy ART: Fast stable learning and categorization of analog patterns by an adaptive resonance system. *Neural Networks*, *4*(6), 759–772.

Christofides, N. (1975). *Graph theory–an algorithmic approach*. New York: Academic Press.

Cichocki, A. & Unbehauen, R. (1993). *Neural networks for optimization and signal processing*. New York: Wiley.

Cottrell, G. W., Munro, P., & Zipser, D. (1987). Learning internal representations from grey-scale images: An example of extensional programming. In: *Ninth Annual Conference of the Cognitive Science Society*, (pp. 462–473 ). Hillsdale, NJ: Lawrence Erlbaum Associates.

Cowan, J. D., Tesauro, G., & Alspector, J. (Eds.) (1994). *Advances in neural information processing systems 6*. San Mateo, CA: Morgan Kaufmann.

Cowan, M. W. (1987). Die Entwicklung des Gehirns (Brain development). In: *Gehirn und Nervensystem (Brain and Nervous System)*, (pp. 100–111 ). Heidelberg: Spektrum der Wissenschaft.

DARPA (Defense Advanced Research Project Agency) (1988). *DARPA neural network study*. Fairfax, Virginia: AFCEA (Armed Forces Communications and Electronics Association), Int. Press.

Desimone, R. & Schein, S. J. (1987). Visual properties of neurons in area V4 of the macaque: Sensitivity to stimulus form. *Journal of Neurophysiology*, *53*(3).

Draper, B., Brodley, C., & Utgoff, P. (1994). Goal–directed classification using linear machine decision trees. *IEEE Transactions on Pattern Analysis and Machine Intelligence*, *16*(9), 888–893.

Duda, R. O. & Hart, P. E. (1973). *Pattern classification and scene analysis*. New York: Wiley.

Feldman, J. A. & Ballard, D. H. (1982). Connectionist models and their properties. *Cognitive Science*, *6*, 205–254.

Fodor, J. A. & Pylyshyn, Z. W. (1988). Connectionism and cognitive architecture: A critical analysis. *Cognition*, *28*, 3–71.

Freund, R., Haberstroh, B., & Bischof, H. (1992). Tools for dynamic network structures: GRAPE grammars. In: *International Joint Conference on Neural Networks 1992*, volume I, (pp. 737–742 ). Piscataway, NJ: IEEE Press.

Friedman, J. (1973). A recursive partitioning decision rule for nonparametric classification. *IEEE Transactions on Computer*, *26*, 404–408.

Frost, V. S., Stiles, J. A., Shanmugan, K. S., & Holtzman, J. C. (1982). A model for radar images and its application to adaptive filtering of multiplicative noise. *IEEE Transactions Pattern Analysis and Machine Intelligence*, *4*, 157–165.

Fukushima, K. (1984a). Neocognitron: Self-organizing neural network model for a mechanism of pattern recognition unaffected by shift in position. *Biological Cybernetics*, *50*, 105–113.

Fukushima, K. (1984b). A neural network model for the mechanism of feature extraction. *Biological Cybernetics*, *50*, 377–384.

Fukushima, K. (1986). A neural network model for selective attention in visual pattern recognition. *Biological Cybernetics*, *55*, 5–15.

Fukushima, K., Miyake, S., & Ito, T. (1983). Neocognitron: A neural network model for a mechanism of visual pattern recognition. *IEEE Transactions on Systems, Man, and Cybernetics*, *13*(3), 826–834.

Geman, S., Bienenstock, E., & Doursat, R. (1992). Neural networks and the bias/variance dilemma. *Neural Computation*, *4*, 1–58.

Geman, S. & Geman, D. (1984). Stochastic relaxation, gibbs distributions, and the Bayesian restoration of images. *IEEE Transactions on Pattern Analysis and Machine Intelligence*, *6*, 721–741.

Giles, L. C. & Maxwell, T. (1987). Learning, invariance, and generalization in high-order neural networks. *Applied Optics*, *26*(23), 4972–4978.

Girosi, F. & Poggio, T. (1990). Networks and the best approximation property. *Biological Cybernetics*, *63*, 169–176.

Goldberg, D. E. (1989). *Genetic algorithms in search, optimization, and machine learning*. Reading, MA: Addision-Wesley.

Gomory, R. E. & Hu, T. C. (1961). Multi-terminal network flows. *Journal of the Society for Industrial and Applied Mathematics*, *9*(4), 551–570.

Grace, A. E. & Spann, M. (1991). A comparision between Fourier–Mellin descriptors and moment based features for invariant object recognition using neural networks. *Pattern Recognition Letters*, *12*, 635–643.

Grosky, W. I. & Jain, R. (1986). A pyramid based approach to segmentation applied to region matching. *IEEE Transactions Pattern Analysis and Machine Intelligence*, *8*(5), 639–650.

Guo, H. & Gelfand, S. (1992). Classification trees with neural network feature extraction. *IEEE Transactions on Neural Networks*, *3*(6), 923–933.

Hancock, P. J. (1988). Data representation in neural nets: an empirical study. In: *Proc. of the 1988 Connectionist Models Summer School*, (pp. 11–20 ). San Mateo, CA: Morgan Kaufmann.

Hanson, S. J., Cowan, J. D., & Giles, C. L. (Eds.) (1993). *Neural information processing systems 5*. San Mateo, CA: Morgan Kaufman.

Haralick, R. M., Lin, C., Lee, J. S. J., & Zhuang, X. (1987). Multi-resolution morphology. In: *Proceedings of the First International Conference on Computer Vision*, (pp. 516–520 ). IEEE Press.

Haralick, R. M. & Shapiro, L. G. (1991). Glossary of computer vision terms. *Pattern Recognition*, *24*, 69–93.

Hartley, R. L. (1984). *Multi–scale models in image analysis*. Unpublished doctoral dissertation, University of Maryland, College Park, Computer Science Center.

Hartmann, G., Kräuter, K. O., Wiemers, H., Seidenberg, E., & Drüe, S (1993). Ein distanz- und orientierungsinvariantes lernfähiges Erkennungssystem für Roboteranwendungen (A distance– and orientation invariant learning and recognition systems for application in robotics). In: *Mustererkennung 1993*, Pöppl, S. J. & Handels, H. (Eds.), (pp. 375–382 ). Berlin: Springer-Verlag.

Haykin, S. (1994). *Neural networks: A comprehensive foundation*. New York: McMillan.

Hebb, D. O. (1949). *The organization of behavior*. New York: Wiley.

Hertz, J., Krough, A., & Palmer, R. G. (1991). *Introduction to the theory of neural computation*. Redwood City, CA: Addison–Wesley.

Hiari, Y. & Tsukui, Y. (1990). Position independent pattern matching by neural networks. *IEEE Transactions System, Man and Cybernetics*, *20*(4), 816–825.

Hinton, G. (1989). Connectionist learning procedures. *Artificial Intelligence*, *40*, 185–234.

Hinton, G. E. (1990). Mapping part-whole hierarchies into connectionist networks. *Artificial Intelligence*, *46*, 47–75.

Hinton, G. E., McClelland, J. L., & Rumelhart, D. E. (1986). Distributed representations. In: *Parallel Distributed Processing*, Rumelhart, D. & McClelland, J. (Eds.), volume 1, (pp. 77–109 ). Cambridge, MA: MIT Press.

Hinton, G. E. & Sejnowski, T. J. (1986). Learning and relearning in Boltzmann machines. In: *Parallel Distributed Processing*, Rumelhart, D. & McClelland, J. (Eds.), volume 1, (pp. 282–317 ). Cambridge, MA: MIT Press.

Hopcroft, J. E. & Ullman, J. D. (1979). *Introduction to automata theory, languages and computations*. Reading, MA: Addison Wesley.

Hopfield, J. J. (1982). Neural networks and physical systems with emergent collective computational abilities. *Proceedings of National Academy of Sciences USA*, *79*, 2554–2558.

Hopfield, J. J. (1984). Neurons with graded response have collective computational properties like those of two-state neurons. *Proceedings of National Academy of Sciences USA*, *82*, 3088–3092.

Hopfield, J. J. & Tank, D. W. (1985). Neural computation of decisions in optimization problems. *Biological Cybernetics*, *52*, 141–152.

Hornik, K. (1991a). Approximation capabilities of multilayer feedforward networks. *Neural Networks*, *4*, 251–257.

Hornik, K. (1991b). Functional learning and approximation in artificial neural networks. *Neural Network World*, *1*, 257–266.

Hornik, K. & Kuan, C. M. (1992). Convergence analysis of local feature extraction algorithms. *Neural Networks*, *5*(2), 229–240.

Hubel, D. H. & Wiesel, T. N. (1986). Die Verarbeitung visueller Information (Processing of visual information). *Heidelberg: Spektrum der Wissenschaft*, *Wahrnehmung und visuelles System (Perception and visual System)*, 36–47.

Jacobs, R. A., Jordan, M. I., & Barto, A. G. (1990). *Task decomposition through competition in a modular connectionist architecture: The what and where vision tasks*. Tech. Rep. No, COINS TR–90–27, University of Massachusetts, Amherst, Department of Computer and Information Science.

Jolion, J. M. & Montanvert, A. (1992). The adaptive pyramid, a framework for 2D image analysis. *Computer Vision, Graphics, and Image Processing: Image Processing*, *55*(3), 339–348.

Jolion, J.-M. & Rosenfeld, A. (1994). *A pyramid framework for early vision*. Boston: Kluwer.

Jordan, M. I. (1986). Attractor dynamics and parallelism in a connectionist sequential machine. In: *Proceedings of 8th Annual Conference of the Cognitive Science Society*, (pp. 531–546 ). Hillsdale NJ: Lawrence Erlbaum Associates.

Kamgar-Parsi, B., Gualtieri, J. A., Devaney, J. E., & Kamgar-Parsi, B. (1990). Clustering with neural networks. *Biological Cybernetics*, *63*, 201–208.

Kamgar-Parsi, B. & Kamgar-Parsi, B. (1992). Hopfield model and optimization problems. In: *Neural Networks for Perception*, Wechsler, H. (Eds.), volume 2, (pp. 94–110 ). Boston: Academic Press.

Kandel, E. R. & Schwartz, J. H. (1985). *Principles of neural science*. New York: Elsevier.

Kass, M., Witkin, A., & Terzopoulos, D. (1987). SNAKES: Active contour models. In: *Proceedings of the First International Conference on Computer Vision*, (pp. 259–268 ). IEEE Press.

Kim, K. & Crawford, M. M. (1991). Adaptive parametric estimation and classification of remotely sensed imagery using a pyramid structure. *IEEE Transactions on Geoscience and Remote Sensing*, *29*(4).

Kirkpatrick, S., Gelatt, C. D., & Vecchi, M. P. (1988). Optimization by simulated annealing. *Science*, *220*, 671–680.

Kohonen, T. (1989). *Self-organization and associative memory*. Berlin: Springer–Verlag.

Kropatsch, W. G. (1986). Curve representations in multiple resolutions. In: *Proceedings of Eighth International Conference on Pattern Recognition*, (pp. 1283–1285 ). Washington: IEEE Computer Society Press.

Kropatsch, W. G. (1987a). Curve representations in multiple resolutions. *Pattern Recognition Letters*, *6*(3), 179–184.

Kropatsch, W. G. (1987b). Elimination von "kleinen" Kurvenstücken in der $2 \times 2/2$ Kurvenpyramide (Elimination of "short" curves in the $2 \times 2/2$ curve pyramid). In: *Mustererkennung 1987*, Paulus, E. (Eds.), Informatik Fachberichte 149, (pp. 156–160 ). Berlin: Springer-Verlag.

Kropatsch, W. G. (1991). *Image pyramids and curves: An overview*. Tech. Rep. No, PRIP-TR-2, Technical University Vienna, Dept. for Pattern Recognition and Image Processing.

Kropatsch, W. G. (1992). Irregular pyramids. In: *Modelling and New Methods in Image Processing and in Geographical Information Systems*, Mandl, P. (Eds.), volume 61 of *OCG-Schriftenreihe*, (pp. 39–50 ). Wien: Oldenbourg–Verlag.

Kropatsch, W. G. & Montanvert, A. (1992). *Irregular pyramids*. Tech. Rep. No, PRIP-TR-5, Technical University Vienna, Department for Pattern Recognition and Image processing.

Kropatsch, W. G., Reither, C., Willersinn, D., & Wlaschitz, G. (1993). The dual irregular pyramid. In: *5th International Conference CAIP'93 on Computer Analysis of Images and Patterns*, (pp. 31–41 ). Berlin: Springer-Verlag.

Kropatsch, W. G. & Willersinn, D. (1992a). Curve representation in different grids. In: *Theoretical Foundations of Computer Vision*, Klette, R. & Kropatsch, W. G. (Eds.), volume 69 of *Mathematical Research*, (pp. 141–154 ). Berlin: Akademie Verlag.

Kropatsch, W. G. & Willersinn, D. (1992b). Representing curves in irregular pyramids. In: *Pattern Recognition 1992*, Bischof, H. & Kropatsch, W. (Eds.), volume 62 of *OCG-Schriftenreihe*, (pp. 233–248 ). Wien: Oldenbourg-Verlag.

Kropatsch, W. G. & Willersinn, D. (1994). Irregular curve pyramids. In: *Shape in Picture, Mathematical Description of Shape in Grey-level Images*, O, Y.-L., Foster, D., Heijmans, H. J., & Meer, P. (Eds.), volume 126 of *NATO-ASI*, (pp. 525–537 ). Berlin: Springer.

Kung, S. Y. & Diamantaras, K. I. (1991). Neural networks for extracting unsymmetric principal components. In: *Neural Networks for Signal Processing, Proceedings of IEEE Workshop*, Juang, B. H. (Eds.), (pp. 50–59 ). Washington: IEEE Press.

Landelius, T. (1993). *Behavior representation by growing a learning tree.* Unpublished doctoral dissertation, Linköping University, Sweden.

Le Cun, Y., Boser, B., Denker, J. S., Henderson, D., Howard, R. E., Hubbard, W., & Jackel, L. D. (1990a). Handwritten digit recognition with a back-propagation network. In: *Advances in Neural Information Processing Systems 2*, Touretzky, D. (Eds.), (pp. 396–404 ). San Mateo, CA: Morgan Kaufmann.

Le Cun, Y., Matan, O., Boser, B., Denker, J. S., Henderson, D., Howard, R. E., Hubbard, W., Jackel, L. D., & Baird, H. S. (1990b). Handwritten ZIP code recognition with multilayer networks. In: *Proceedings of the 10.International Conference on Pattern Recognition*, (pp. 35–40 ). Washington: IEEE Computer Society Press.

Le Cun, Y., Matan, O., Boser, B., Denker, J. S., Henderson, D., Howard, R. E., Hubbard, W., Jackel, L. D., & Baird, H. S. (1990c). *Handwritten ZIP code recognition with multilayer networks.* Tech. Rep. No, 38543, AT&T Bell Laboratories.

Le Cun, Y. L. (1986). Learning processes in an asymetric threshold network. In: *Disordered Systems and Biological Organization*, Bienenstock, E., Fogelman-Souli, F., & Weisbruch, G. (Eds.), (pp. 233–240 ). Berlin: Springer–Verlag.

Le Cun, Y. L., Denker, J. S., & Solla, S. A. (1988). Optimal brain damage. In: *Advances in Neural Information Processing Systems 2*, Touretzky, D. S. (Eds.), (pp. 598–605 ). San Mateo, CA: Morgan Kaufmann.

Lee, J. S. (1981). Refined filtering of image noise using local statistics. *Computer Vision, Graphics, and Image Processing*, *15*, 380–389.

Lehky, S. R. & Sejnowski, T. J. (1988). Network model of shape-from-shading: Neural functions arise from both the receptive and projective field. *Nature*, *333*, 452–454.

Levialdi, S. (1986). Programming image processing machines. In: *Pyramidal Systems for Image Processing and Computer Vision*, Levialdi, S. & Cantoni, V. (Eds.), volume F25 of *NATO ASI Series*, (pp. 311–328 ). Berlin: Springer-Verlag.

Linhart, G. & Dorffner, G. (1992). A self-learning visual pattern explorer and recognizer using high order neural networks. In: *Proceedings of the International Joint Conference on Neural Networks*, volume III, (pp. 705–710 ). Baltimore: IEEE Computer Society Press.

Linsker, R. (1986). From basic network principles to neural architectures (series). *Proceedings of National Academy of Sciences USA*, *83*, 7508–7512, 8390–8394, 8779–8783.

Linsker, R. (1988). Self-organization in a perceptual network. *IEEE Computer*, *21*, 105–117.

Lippmann, R. L. (1987). An introduction to computing with neural nets. *IEEE Acoustics Speech and Signal Processing Magazine*, *4*, 4–22.

Lippmann, R. P., Moody, J. E., & Touretzky, D. S. (Eds.) (1991). *Advances in neural information processing systems*, volume 3. San Mateo, CA: Morgan Kaufmann.

Mallat, S. G. (1989). A theory for multiresolution signal decomposition: The wavelet representation. *IEEE Transactions on Pattern Analysis and Machine Intelligence*, *11*(7), 674–693.

Marr, D. (1982). *Vision*. New York: Freeman.

Masters, T. (1993). *Practical neural network recipes in C++*. Boston: Academic Press.

Mayer, H. & Kropatsch, W. G. (1989a). Progressive Bildübertragung mit der $3 \times 3/2$ Pyramide (Progressive image transmission with the $3 \times 3/2$ pyramid). In: *Informatik Fachberichte 219: Mustererkennung 1989*, Burkhardt, H., Höhne, K. H., & Neumann, B. (Eds.), (pp. 160–167 ). Berlin: Springer–Verlag.

Mayer, H. F. & Kropatsch, W. G. (1989b). Kompakte Bildkodierung mit der $3 \times 3/2$ Pyramide (Compact image coding with the $3 \times 3/2$ pyramid). In: *Wissensbasierte Mustererkennung*, Pinz, A. (Eds.), OCG-Schriftenreihe, Österr. Arbeitsgemeinschaft für Mustererkennung, (pp. 195–210 ). Wien: Oldenbourg-Verlag. Band 49.

McClelland, J. L. & Rumelhart, D. E. (1986). *Parallel distributed processing: Exploration in the microstructure of cognition: Psychological and biological models*. Cambridge, MA: MIT Press.

McClelland, J. L. & Rumelhart, D. E. (1988). *Explorations in parallel distributed processing*. Cambridge MA: MIT Press.

McEliece, R. J., Posner, E. C., & Rodemich, E. R. (1987). The capacity of the Hopfield associative memory. *IEEE Transactions on Information Theory*, *33*, 461–482.

Meer, P. (1989). Stochastic image pyramids. *Computer Vision, Graphics, and Image Processing*, *45*(3), 269–294.

Meer, P., Sher, C. A., & Rosenfeld, A. (1988). *The chain pyramid: Hierarchical contour processing*. Tech. Rep. No, TR-2072, University of Maryland, College Park, Computer Science Center.

Mel, B. W. & Omohundro, S. M. (1991). How receptive field parameters affect neural learning. In: *Advances in Neural Information Processing 3*, Lippmann, R. P., Moody, J. E., & Tourtetzky, D. S. (Eds.), (pp. 757–763 ). San Mateo, CA: Morgan Kaufmann.

Milanese, R. (1990). *Focus of attention in human vision: a survey*. Tech. Rep. No, 90.03, University of Geneva.

Milanese, R. (1992). A relaxation network for feature driven visual attention. In: *SPIE Conference on Neural and Stochastic methods in Image and Signal Processing, San Diego*, Chen, S.-S. (Eds.), volume SPIE Vol. 1766, (pp. 542–532 ). Bellingham, WA: SPIE-The International Society for Optical Engineering.

Milanese, R. (1994). *Detecting salient regions in an image: From biological evidence to computer implementation.* Unpublished doctoral dissertation, University of Geneva, Department of Computer Science.

Montanvert, A. & Bertolino, P. (1992). Irregular pyramids for parallel image segmentation. In: *Pattern Recognition 1992*, Bischof, H. & Kropatsch, W. G. (Eds.), volume OCG-Schriftenreihe 62, (pp. 13–34 ). Wien: Oldenbourg–Verlag.

Moody, J. E., Hanson, S. J., & Lippmann, R. P. (Eds.) (1992). *Advances in neural information processing systems 4.* San Mateo, CA: Morgan Kaufmann.

Moran, J. & Desimone, R. (1985). Selective attention gates visual processing in the extrastriate cortex. *Science*, *229*, 782–784.

Mozer, M. C. & Smolensky, P. (1988). Skeletonization: A technique for trimming the fat from a network via relevance assessment. In: *Advances in Neural Information Processing Systems*, Touretzky, D. S. (Eds.), volume 1. San Mateo, CA: Morgan Kaufmann.

Murthy, S. K., Kasif, S., & Salzberg, S. (1994). A system for induction of oblique decision trees. *Journal of Artificial Intelligence Research*, *2*, 1–32.

Nagao, M. & Matsuyama, T. (1979). Edge preserving smoothing. *Computer Graphics and Image Processing*, *9*, 394–407.

Oja, E. (1982). A simplified neuron model as a principal component analyzer. *Journal of Mathematical Biology*, *15*, 267–273.

Oja, E. (1989). Neural networks, principal components, and subspaces. *International Journal of Neural Systems*, *1*(1), 61–68.

Olshausen, B., Anderson, D., & Van Essen, C. (1992). *A neural model of visual attention and invariant pattern recognition.* Tech. Rep. No, CNS Memo 18, Caltech Institute of Technology.

Omohundro, S. M. (1989a). *Five balltree construction algorithms.* Tech. Rep. No, TR-89-063, Berkeley, International Computer Science Inst.

Omohundro, S. M. (1989b). *Geometric learning algorithms.* Tech. Rep. No, TR-89-041, Berkeley, International Computer Science Institute.

Omohundro, S. M. (1990). *The delaunay triangulation and function learning.* Tech. Rep. No, TR-90-001, Berkeley, International Computer Science Institute.

Omohundro, S. M. (1991). Bumptrees for efficient function, constraint, and classification learning. In: *Advances in Neural Information Processing*, Lippmann, R. P., Moody, J. E., & Tourtetzky, D. S. (Eds.), volume III, (pp. 693–699 ). San Mateo, CA: Morgan Kaufmann.

Omohundro, S. M. (1992). Best-first model merging for dynamic learning and recognition. In: *Advances in Neural Information Processing*, Moody, J. E., Hanson, S. J., & Lippmann, R. P. (Eds.), volume IV, (pp. 958–965 ). San Mateo, CA: Morgan Kaufmann.

Pao, Y. H. (1989). *Adaptive pattern recognition and neural networks.* Reading, MA: Addison-Wesley.

Park, D. J., Park, R. H., Lee, S. U., & Choi, J. S. (1994). Hierarchical edge detection using the bidirectional information in edge pyramids. *Pattern Recognition Letters*, *15*, 65–75.

Parker, D. B. (1985). *Learning-logic.* Tech. Rep. No, 47, MIT, Sloan School of Management,Cambridge, MA.

Peterson, C. & Anderson, J. A. (1987). A mean field learning algorithm for neural networks. *Complex Systems*, *1*, 995–1019.

Phatak, D. S., Choi, H., & Koren, I. (1993). Construction of minimal $n-2-n$ encoder for any $n$. *Neural Computation*, *5*(5), 783–794.

Pineda, F. J. (1987). Generalization of back-propagation to recurrent neural networks. *Physical Review Letters*, *59*, 2229–2232.

Pinz, A. & Bischof, H. (1990). *The IVF weight-visualization diagrams (WV-Diagrams).* Tech. Rep. No, 1, Institut für Vermessungswesen, Vienna, Univ. of Agriculture.

Poggio, T. & Girosi, F. (1990). Networks for approximation and learning. *Proceedings of the IEEE*, *78*(9), 1481–1497.

Pollack, J. B. (1990). Recursive distributed representations. *Artificial Intelligence*, *46*, 77–105.

Qian, N. & Sejnowski, T. J. (1988). Predicting the secondary structure of globular proteins using neural network models. *Journal of Molecular Biology*, *202*, 865–884.

Quinlan, J. (1986). Induction of decision trees. *Machine Learning*, *1*, 81–106.

Ripley, B. D. (1993). Statistical aspects of neural networks. In: *Network and Chaos – Statistical and Probabilistic Aspects*, Barndorff-Nielsen, O. E., Jensen, J. L., & Kendall, W. S. (Eds.). Chapman & Hall.

Ritter, G. X., Li, D., & Wilson, J. N. (1989). Image algebra and its relationships to neural networks. In: *Proceedings of SPIE Technical Symposion Southeast on Optics, Electronics-Optics, and Sensors*, (pp. 90–101 ). Bellingham, WA: SPIE-The International Society for Optical Engineering.

Rojer, A. & Schwartz, E. L. (1990). Design of a space variant sensor having a complex log geometry. In: *Proceedings of 10th International Conference on Pattern Recognition*, volume II, (pp. 278–285 ). Washington: IEEE Computer Society Press.

Rosenblatt, F. (1962). *Principles of neurodynamics*. New York: Spartan.

Rosenfeld, A. (Eds.) (1984). *Multiresolution image processing and analysis*. Berlin: Springer–Verlag.

Rosenfeld, A. (1985a). *Arc colorings, partial path groups, and parallel graph contractions*. Tech. Rep. No, TR-1524, University of Maryland, Computer Science Center.

Rosenfeld, A. (1985b). The prism machine: An alternative to the pyramid. *Journal of Parallel Distributed Computing*, *2*, 404–411.

Rosenfeld, A. (1987). Recognizing unexpected objects: A proposed approach. *International Journal of Pattern Recognition and Artificial Intelligence*, *1*(1), 71–84.

Rota, G. C. (1969). *Combinatorial analysis, the mathematical sciences: A collection of essays*. Cambridge, MA: MIT Press.

Roth, M. W. (1990). Survey of neural network technology for automatic target recognition. *IEEE Transactions on Neural Networks*, *1*(1), 28–43.

Rumelhart, D. E., Hinton, G. E., & Williams, R. J. (1986). Learning representations by back-propagating errors. *Nature*, *323*, 533–536.

Rumelhart, D. E. & McClelland, J. A. (1986). *Parallel distributed processing 1*. Cambridge, MA: MIT Press.

Rummelhart, D. E. & Zipser, D. (1985). Feature discovery by competetive learning. *Cognitive Science*, *9*, 75–112.

Saha, A., Christian, J., Tang, D. S., & Wu, C.-L. (1991). Oriented non-radial basis functions for image coding and analysis. In: *Advances in Neural Information Processing 3*, Lippmann, R. P., Moody, J. E., & Tourtetzky, D. S. (Eds.), (pp. 728–734 ). San Mateo, CA: Morgan Kaufmann.

Sahoo, P. K., Soltani, S., & Wang, A. K. C. (1988). A survey of thresholding techniques. *Computer Vision, Graphics, and Image Processing: Image Processing*, *41*, 233–260.

Samet, H. (1990). *The design and analysis of spatial data structures*. Reading, MA: Addison–Wesley.

Sanger, T. D. (1989). Optimal unsupervised learning in a single-layer linear feedforward neural network. *Neural Networks*, *2*, 433–459.

Sanger, T. D. (1991). Basis-function trees as a generalization of local variable selection methods. In: *Advances in Neural Information Processing 3*, Lippmann, R. P., Moody, J. E., & Tourtetzky, D. S. (Eds.), (pp. 700–706 ). San Mateo, CA: Morgan Kaufmann.

Schmidhuber, J. & Prelinger, D. (1993). Discovering predictable classifications. *Neural Computation*, *5*(4), 625–635.

Schrift, A., Zeevi, Y. Y., & Porat, M. (1988). Pyramidal edge detection and image representation. In: *SPIE Proceedings Visual Communications and Image Processing '88*, volume SPIE–1001, (pp. 529–536 ). Bellingham, WA: SPIE-The International Society for Optical Engineering.

Schweizer, L., Parladori, G., Sicranza, G. L., & Marsi, S. (1991). A fully neural approach to image compression. In: *Artificial Neural Networks*, Kohonen, T., Mäkissara, K., Simula, O., & Kangas, J. (Eds.), volume I, (pp. 815–820 ). Berlin: Springer–Verlag.

Seibert, M. & Waxman, A. (1989). Spreading activation layers, visual saccades, and invariant representations for neural pattern recognition systems. *Neural Networks*, *2*, 9–27.

Serra, J. (1982). *Image analysis and mathematical morphology*. New York: Academic Press.

Shrivastava, Y., Dasgupta, S., & Reddy, S. M. (1992). Guaranteed convergence in a class of Hopfield networks. *IEEE Transactions on Neural Networks*, *3*(6), 951–961.

Singer, W. (1986). Hirnentwicklung und Umwelt (Brain development and environment). In: *Wahrnehmung und visuelles System (Perception and visual system)*, (pp. 186–199 ). Spektrum der Wissenschaft, Heidelberg.

Smolensky, P. (1990). Tensor product variable binding and the representation of symbolic structures in connectionist systems. *Artificial Intelligence*, *46*, 159–216.

Sorenson, H. & Alspach, D. (1971). Recursive Bayesian estimation using Gaussian sums. *Automatica*, *7*, 465–479.

Soukoulis, C. M., Levin, K., & Grest, G. S. (1983). Irreversibility and metastability in spin-glasses. I. Ising model. *Physical Review B*, *28*, 1495–1509.

Tanimoto, S. L. (1986). Paradigms for pyramid machine algorithms. In: *Pyramidal Systems for Image Processing and Computer Vision*, Levialdi, S. & Cantoni, V. (Eds.), volume F25 of *NATO ASI Series*, (pp. 173–194 ). Berlin: Springer-Verlag.

Thurner, E. (1994). *Modulare Neurale Systeme (Modular neural systems)*. Tech. Rep. No, PRIP-TR-31, Technical University Vienna, Department for Pattern Recognition and Image Processing.

Thurner, E. & Bischof, H. (1994). Modulare neuronale Architekturen in der Mustererkennung (Modular neural architectures for pattern recognition). In: *Mustererkennung 1994*, Kropatsch, W. G. & Bischof, H. (Eds.), volume 5 of *Informatik Xpress*, (pp. 566–574 ). Berlin: Springer Verlag.

Touretzky, D. S. (Eds.) (1989). *Advances in neural information processing systems 1*. San Mateo, CA: Morgan Kaufmann.

Touretzky, D. S. (Eds.) (1990). *Advances in neural information processing systems 2*. San Mateo, CA: Morgan Kaufmann.

Tsotsos, J. K. (1987). A "complexity level" analysis of vision. In: *Proceedings of the First International Conference on Computer Vision*, (pp. 346–355 ). London, England: IEEE Press.

Tsotsos, J. K. (1990). Analyzing vision at the complexity level. *Behavioral and Brain Sciences*, *13*(3), 423–469.

Tsotsos, J. K. (1991). *Localizing stimuli in a sensory field using an inhibitory attentional beam.* Tech. Rep. No, RBCV-TR-91-37, University of Toronto.

Tsotsos, J. K. (1992). On the relative complexity of active vs. passive visual search. *International Journal of Computer Vision*, *7*(2), 127–141.

Uhr, L. (Eds.) (1987). *Parallel computer vision.* Boston: Academic Press.

Utgoff, P. & Brodley, C. (1990). An incremental method for finding multivariate splits for decision trees. In: *Proceedings of the Seventh International Conference on Machine Learning*, (pp. 58–65 ). Austin TX: Morgan Kaufman.

Wagner, R. (1991). *Verfolgung der Siedlungsentwicklung mit Landsat TM Bilddaten (Settlement development with Landsat TM data).* Unpublished doctoral dissertation, Inst. f. Surveying and Remote Sensing, University of Natural Resources, Vienna.

Wallace, R. S., Ong, P.-W., Bederson, B. B., & Schwartz, E. L. (1993). *Space variant image processing.* Tech. Rep. No, 633, New York University.

Watkins, C. (1989). *Learning with delayed rewards.* Unpublished doctoral dissertation, Cambridge University, Psychology Department.

Werbos, P. J. (1974). *Beyond regression: New tools for prediction and analysis in the behavioral sciences.* Unpublished doctoral dissertation, Harvard University.

White, H. (1990). Connectionist nonparametric regression: Multilayer networks can learn arbitrary mappings. *Neural Networks*, *3*, 535–549.

Widrow, B. & Stearns, S. D. (1985). *Adaptive signal processing.* New York: Prentice-Hall.

Wilson, S. S. (1989). Vector morphology and iconic neural networks. *IEEE Transactions on Systems, Man, and Cybernetics*, *19*(6), 1636–1644.

Wu, Z. & Leahy, R. (1993). An optimal graph theoretic approach to data clustering: Theory and its application to image segmentation. *IEEE Transactions on Pattern Recognition and Machine Intelligence*, *15*(11), 1101–1113.

Xu, L. (1993). Least mean square error reconstruction principle for self-organizing neural-nets. *Neural Networks*, *6*, 627–648.

Xu, L., Krzyzak, A., & Yuille, A. (1994). On radial basis function nets and kernel regression: Statistical consistency, convergence rates, and receptive field size. *Neural Networks*, *7*(4), 609–628.

Yamaguchi, T. (1990). *The image pattern recognition by a neural network*. Unpublished doctoral dissertation, Johannes Kepler University, Linz.

Yamaguchi, T. & Kropatsch, W. G. (1989). A vision by neural network or by pyramid. In: *Proceedings of the 6th Scandinavian Conference on Image Analysis*, (pp. 104–111 ). Berlin: Springer-Verlag.

Yamaguchi, T. & Kropatsch, W. G. (1990). Distortion-tolerance curve of the Neocognitron with various structures, including pyramid. In: *Proceedings 10th International Conference on Pattern Recognition*, volume I, (pp. 918–922 ). Atlantic City, NJ: IEEE Computer Society Press.

Yuille, A. L., Kammen, D. L., & Cohen, D. S. (1989). Quadrature and the development of orientation selective cortical cells by Hebb rules. *Biological Cybernetics*, *61*, 183–194.

Yuille, A. L. & Poggio, T. A. (1986). Scaling theorems for zero crossings. *IEEE Transactions on Pattern Analysis and Machine Intelligence*, *8*(1), 15–25.

Zeki, S. (1993). *A vision of the brain*. Oxford: Blackwell.

Ziv, J. & Lempel, A. (1977). A universal algorithm for sequential data compression. *IEEE Transactions on Information Theory*, *23*(5), 337–343.

# Author Index

# Subject Index